The Story of L Street Power Station 1898–2006

By Gilmore G. Cooke

gilcooke@ieee.org

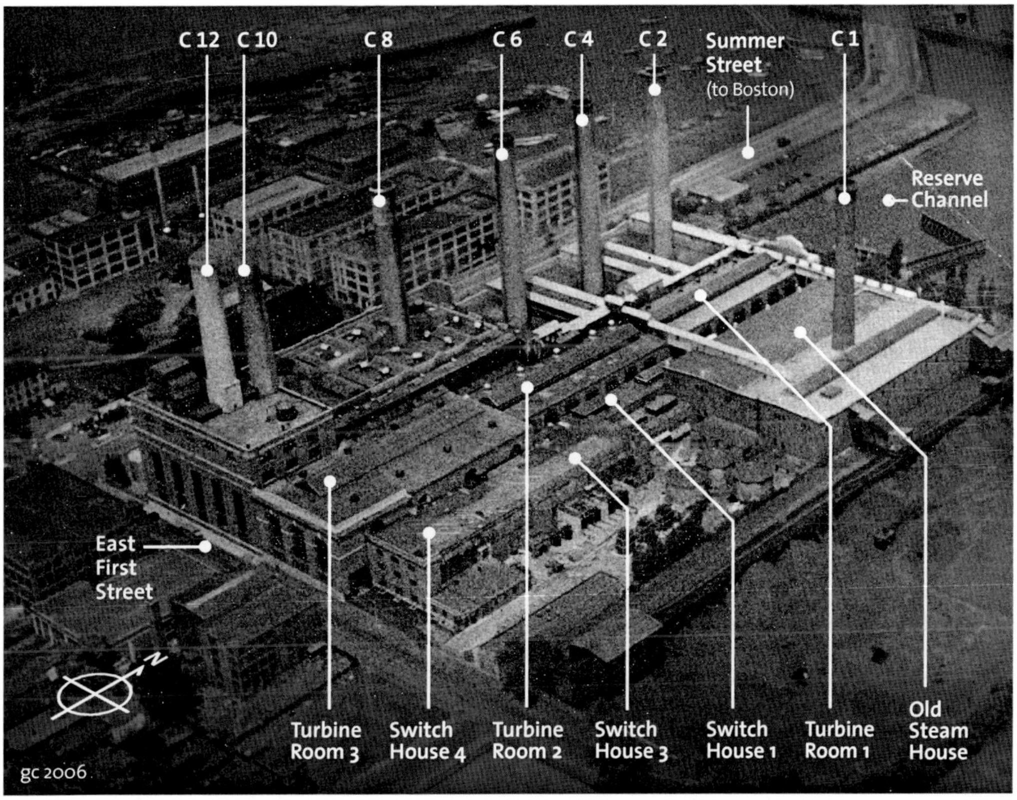

View of Boston Edison's L Street Power Generating Station from a photograph taken in the 1950s. The view is looking northward towards the City of Boston. Summer Street, originally called L Street, is visible crossing the Reserve Channel in South Boston. L Street's seven chimneys were operating at the time this aerial view was taken, with the exception of chimney C1, which was decommissioned in 1935. These chimneys formed part of four distinct and identifiable power stations built progressively starting in 1898. In the early sixties, the New Boston Generating Station was built, completely reconfiguring the facility. This book presents the evolution of this power generating site in five chapters or parts, beginning in 1898 and extending to 2006.

First Published 2007
by Exelon Corporation
200 Exelon Way
Kennett, Pennsylvania

Designed by Rector Communications, Inc., Jim Yost, Designer,
www.rector.com
gilcooke@ieee.org

ISBN: 978-1-7332378-2-6

The Story of the L Street Power Station 1898–2006

By Gilmore G Cooke

Content Summary:

Acknowledgements, Abbreviations and Names — IV

Brief Chronology — V

Part 1 — 1

Part 1 describes the history of the steam power generating plant associated with chimney number C1. It served boilers numbered #1 through #12. The old engine house consisted of six reciprocating steam engine-generators. A description of the 1898 power system is given together with an account of its operating history.

Part 2 — 13

Part 2 describes the historical development of the steam power generating system associated with chimneys C2, C4, and C6. These chimneys served boilers numbered #13 through #60, located in Boiler Room 1 and 2. The system included a 200 psi steam supply, eight vertical turbine generators numbered T #1 through T #8 and auxiliaries. The original power system and the electrical system are described, including a brief account of the station's history, and the organization of the operating staff.

Part 3 — 51

Part 3 describes the historical development of the steam power generating system associated with chimneys C8 and C10. These two chimneys served twelve boilers, numbered #61 through #72, located in Boiler Room 3. The system included a 300 psi steam supply; three horizontal turbine generators numbered T #9, T #10 and T #11, and auxiliaries. A description of the original power system is given, with a brief account of its history and operations.

Part 4 — 75

Part 4 is the historical development of the steam power generating system associated with chimney C12. This chimney served three boilers, numbered #74, #75 and #76, shoehorned into Boiler Room 3. The system included a high-pressure 1,400 psi steam supply, two "topping" turbine generators, numbered T #1201 and T #1202, and auxiliaries. A description of the steam power system is given, with a brief account of its history and operations. The station's large frequency converter is discussed in this section.

Part 5 — 89

Part 5 is a brief history of New Boston Generating Station.

References — 104

Acknowledgements

I wish to thank the Exelon Generation Company and the employees of the New Boston Generating Station for sharing their time, knowledge and resources towards my research. Special thanks to John Castro for his encouragement and enthusiastic support throughout this history project. Thanks to Mary Rucci for getting the history published. Thanks also to David O'Brien, William Dougherty, John Ayvazian, Dave Belanger and Louis Clark for escorting me around the L Street site and for sharing documents and drawings. The L Street story would be incomplete without the help and cooperation received from the following organizations and staff: Aaron Schmidt of the Boston Public Library; Harvard University Baker Library, Historical Collections; Melissa Feiden, Moses Carr, and Cassandra Fox of MIT's Library Storage Annex; Chris Hunter and Tony Scalise, Schenectady Museum, General Electric Collection.

Gilmore G. Cooke
February 2007
gilcooke@ieee.org

Photographs

All photographs and images are from the author's collection, except those noted otherwise, where permission was given to the author for their use. Savaria photographs are used with permission of Edward Jr. and Suzanne Savaria Photography Services, Philadelphia, PA.

Abbreviations

AC ______ alternating current
DC ______ direct current
°F ______ degrees Fahrenheit
gpm ______ gallons per minute
hp ______ horsepower
in. ______ inch
kv ______ kilovolt
kva ______ kilovolt amperes
kw ______ kilowatt
kwh ______ kilowatt hour
lb. ______ pound
psi ______ pounds per square inch
rpm ______ revolution per minute

Names

B&W ______ Babcock and Wilcox Company
EEI ______ Edison Electric Illuminating Company of Boston
GE ______ General Electric Company

Brief Chronology

1898: Boston Electric Light Co. builds power station on L Street.

1903: EEI builds new turbine power plant alongside original station.

1904: Turbine #1: in service October 17.

1904: Turbine #2: started October 21.

1906: Turbine #3: rolled with steam November 28. Unit begins service August 27, 1907.

1907: Turbine #4: online December 1.

1908: Turbine #5: started September 6.

1911: Turbine #6: started November 26.

1912: Turbine #7: first load September 14.

1914: Turbine #8: online October 20.

1920: Turbine #9: online September 28.

1921: Turbine #10: carrying 20,500 kw April 21.

1924: Turbine #11: at load August 11.

1939: Turbine #1201: in service c1939.

1942: Turbine #1202: in service c1942.

1965: New Boston Unit 1: commercial date August 11.

1967: New Boston Unit 2: commercial date July 15.

Turbine #1 was retired in 1965. The vertical steam turbines were eventually placed in reserve status. All were line tested everyday till 1968 or 1969. The last of the great verticals, Turbine 8, was taken off line in August 1967.

The 300 psi boilers were shutdown around 1973.

Turbine #9, 10 & 11 remained in reserve until 1974.

The 1,400 psi steam system remained in service until approximately 1978, supplying steam to Boston customers.

Turbine #1201 stayed in service exhausting 300 psi steam till 1978.

Turbine #1202 in service till 1974, then was dismantled and removed.

Aerial view of the New Boston Generating Station with the City of Boston in the background. Photograph taken in the late 1990s by Boston Edison.

Part 1 | History of the steam power generating plant associated with chimney number C1.

Figure 1-0 – The subject of Part 1 is the steam power generating system associated with chimney number C1, on the left side of this 1904 photograph. Chimney C1, the boiler house and the steam engine room were built by Boston Electric Light Company in 1898. It was later purchased by Edison Electric Illuminating Company and taken over February 1, 1902. The latter erected the more handsome building on the right, to be discussed later in Part 2. (Boston Edison Company Collection, Baker Library, Harvard Business School).

Early History

In the late 1890s, Boston Electric Light Company had a growing street lighting and power business in Boston. The electric company's central generating station was located on Gilbert Place, on the opposite side of the street where South Station is today. Boston Electric Light Company had to move their power plant because the railroad company was going to build a large railroad terminal, now South Station, on Summer Street. Fortunately, the electric light company was able to find a better site in South Boston, close to Boston Harbor, and proceeded to build their new generating station on L Street.

The site was perfect because there was plenty of cooling water available for condensers. It was also ideal for mooring and unloading coal schooners. Furthermore, the property had a large outdoor area for storing coal. The power station was constructed in 1898. In 1902, Edison Electric Illuminating Company of Boston (EEI) bought out the company and took over control of the property on L Street.

Design of the Original Station

The 1898 power station was state-of-the-art with its twelve Babcock and Wilcox (B&W) steam boilers, six vertical cross-compound McIntosh & Seymour steam engines, and six General Electric Company (GE) alternating current generators. Generators were rated 1,500 kw, three phases, 2,300 volt, 60 cycles. These generators primarily supplied a series of synchronous 200 hp AC motors mechanically coupled to individual Brush arc lighting generators, used for street lights in Boston.

Architecturally, the building exterior was extremely plain, made of bricks on rough stone masonry foundations. The chimney was an immense brick structure, 225 feet high, with a large octagonal base. The interior space, however, was more impressive. The engine room was designed as a show place for visitors. Interior walls were white enameled bricks. The underside of the roof was made of hollow tile plastered with white cement. Windows had glazed glass that prevented direct sunlight from entering into the room, giving diffused illumination. The floor in the engine room was made of terrazzo mosaic. Mechanical and electrical machines were painted a dark maroon. The place was well lit by thirty-two enclosed arc lights evenly spaced around the room.

Each of the twelve B&W boilers were rated 550 hp and produced 160 psi steam. They were equipped with Hawley downdraft furnaces and were hand fired. The chimney had an inside diameter of 11 feet, 6 inches, custom designed to give the furnaces a natural draft. City water was used to supply the boilers. The feed pump installation consisted of three 8-in. x 10-in. motor-driven Gould's triplex pumps. Pump speed was 32 rpm and each unit had a capacity of 208 gpm against 160 psi boiler pressure. There were twelve vertical National feed water heaters with 750 sq. ft. of heating surface. One heater per engine unit took in exhaust steam and a second was piped to receive the pump exhaust. Feed water entered the boilers at about 190 °F. Each engine was equipped with a Blake vertical twin jet condenser and air pump. A 4,500 ton coal bunker was installed above the boilers.

Brush arc generators were located in the engine generator room. Each arc generator was rated 66 amperes, and was direct-connected in pairs to 200 hp, 2,300 volt, three-phase synchronous motors. There was one 150 kw, 2,200 volt to 550 volt motor-generator set for supplying electricity to a motor-driven feed pump and to streetcar lines of the Boston Elevated Railway Company. The engine room also housed a 75 kw motor-driven exciter, and a 50 kw steam engine-driven exciter. These two 125 volt DC exciters were later connected with other sources in the adjoining turbine station to establish a common excitation power supply. There was also a large storage battery located in a small building at the back of the station.

The power station had two switchboards, one above the other: one on the engine room floor for arc light circuits, and the other on a raised gallery for controlling generators, exciters, motor generators, and 2,300 volt service.

Coal Handling

The company burned New River coal from West Virginia. Approximately 90,000 tons was consumed in 1909. The site had ample space to store over six months' supply of coal. The coal field had a capacity of 70,000 tons. Coal unloading equipment was designed and supplied by CW Hunt Company. The hoisting apparatus consisted of a movable unloading tower with a boom and clamshell bucket overhanging the coal vessel. Coal was screened before being dumped onto the conveyor. The conveyor ran along the entire length of the property. The crusher and conveyor each had a 15 hp induction motor. The hoisting crane had a 500 volt motor with Ward Leonard controls. Two operators manned the 1898 coal unloading tower. It was equipped with a one ton bucket making one trip per minute.

Later, a second tower was erected by EEI for their major turbine expansion. This tower was manned by only one operator and could unload coal from either side of the pier. It was equipped with a one and a half ton bucket, making three trips per minute. A Brown hoist gantry crane with cantilever extension was added for moving and distributing coal over the entire coal yard.

EEI's department of standardization and tests continuously monitored coal. Samples were taken from barge loads and from the boiler rooms. Coal used in the engine station boiler room was recorded by checking car weights as fired; that used in the turbine station boiler rooms was weighed by hopper scales located above the stoker spouts. The total weight of coal used by each division of the plant was carefully tabulated and sent daily to the office of the superintendent of generation.

The power company stockpiled large quantities of coal to guard against shutdown of coal shipments or labor strikes. Coal was purchased and transported during the summer months when prices were lower.

1912 Inventory of Station Equipment

By 1912, the inventory of operational equipment housed in the original steam engine station included:

Boilers:

— Twelve 550 hp B&W boilers.

— Total boiler capabilities: 6,600 hp.

— Type of Grates: Hawley down draft.

— Steam Pressure: 160 psi saturated steam.

Prime Movers:

— Six 2,500 hp, 120 rpm, vertical cross compound McIntosh & Seymour engines, each direct connected to one 1,500 kw, 2,300 volt, 60 cycles, 3-phase GE generators.

Auxiliary Apparatus:

- — Twenty one 200 hp, 2,250 volt, 60 cycles, 3-phase synchronous motors, direct connected to 41 Brush arc generators.
- — Two 500 kw induction motor-generators rated 2,250 volt AC to 550 volt DC.
- — One 150 kw synchronous motor-generator 3-unit set, 2,250 volt AC to 550 volts DC and 2,250 volts AC.
- — 125 volt DC storage batteries manufactured by Electric Storage Battery Company, having a one-hour capacity of 153 kw.
- — Three 75 kw synchronous motor-generator exciters, 2,250 volts AC to 125 volt DC.
- — One 75 hp Ames engine direct connected to a 50 kw, 125 volt DC generator.

Decommissioning of the Equipment

Few details are available on the decommissioning of original Boston Electric Light Company equipment. We know that money was authorized during the depression for the removal of obsolete boilers and steam engines. The space was subsequently used for machine shops and storing supplies.

On November 4, 1935, the demolition of the twelve original boilers was approved. On June 22, 1936, money was approved for the removal of the six reciprocating steam engines. Salvage value was recorded at $10,455; replacement value was calculated at $340,486. By 1937, EEI began phasing out Brush arc lighting machines from service.

Photographs and Illustrations

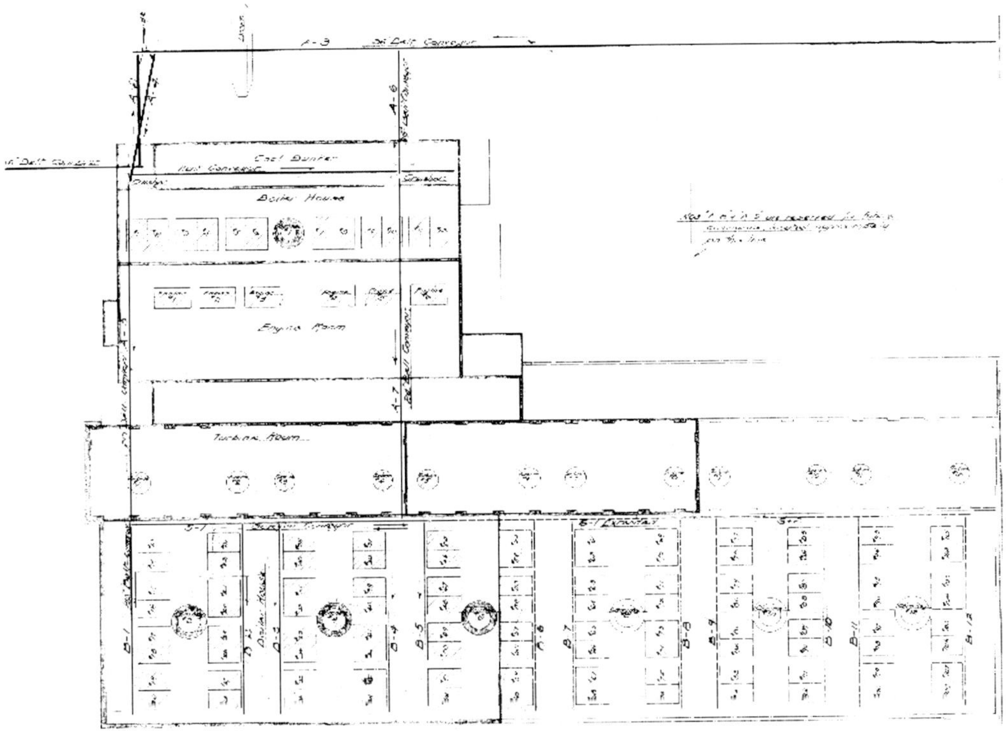

Figure 1-1 – The new property, acquired in 1902, became EEI's fourth electric generating station. Known as Station 4, the name L Street was more commonly used. The company's engineering department established the equipment numbering scheme for station equipment. Boilers were numbered #1 through #12; steam engines were numbered #1 through #6. Chimney C1 appears on the layout drawing. Subsequent chimneys on the turbine side of the facility were given even numbers, C2, C4, and so on. Heavy lines on the plans were part of the coal conveyor system transporting coal from storage, over on rooftops, to coal bunkers in the turbine plant.

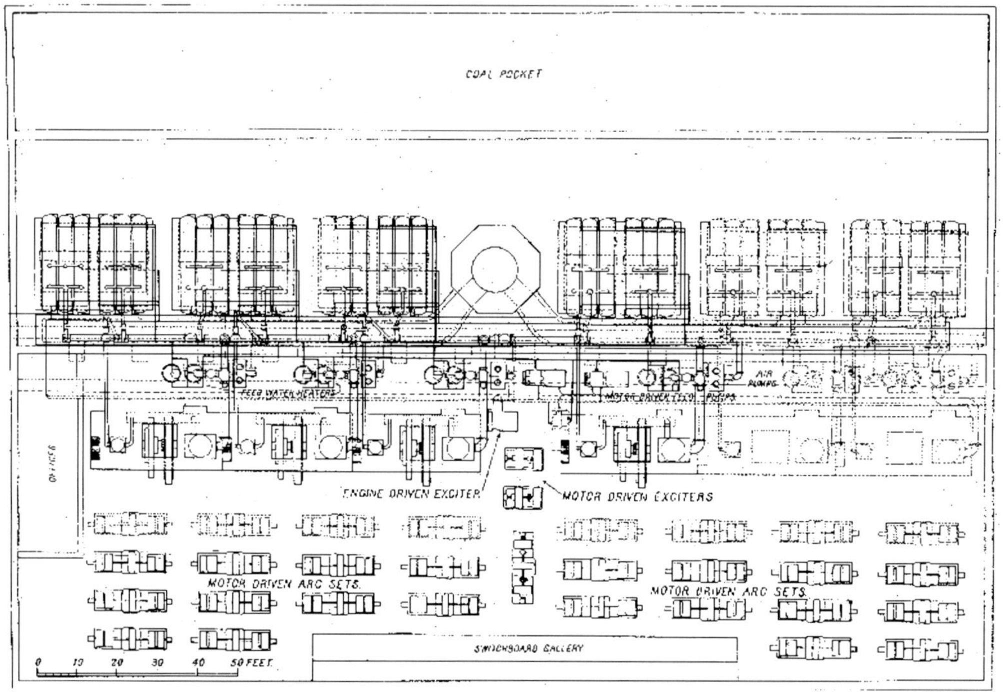

Figure 1-2 – Plan of Boston Electric Light Company central generating station was featured in Electrical World, May 27, 1899. The switchboard gallery on the bottom of the figure still exists. The switchboard is still energized and in use today for 550 volt AC house loads.

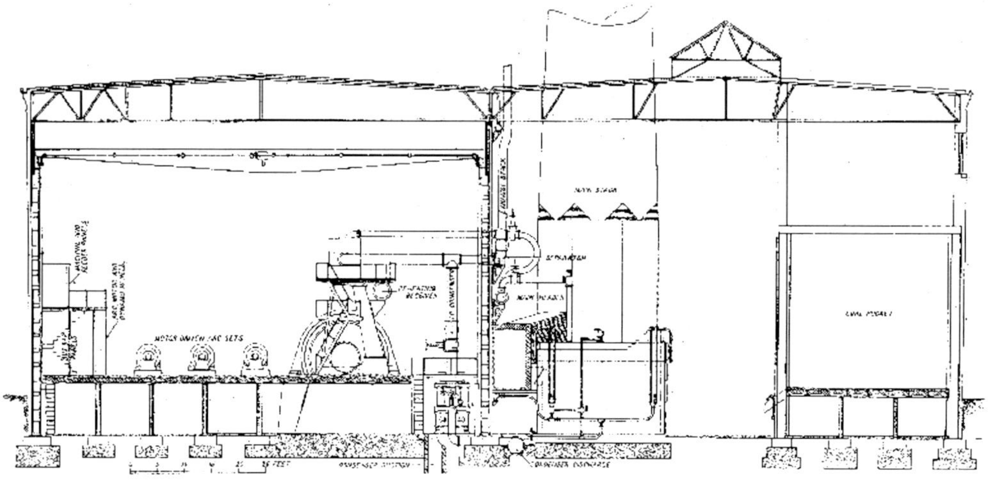

Figure 1-3 – Cross section of the original 1898 station.

Figure 1-4 – The original coal unloading system consisted of a single coal tower (right) and conveying equipment at the end of a pier, running 150 feet into Boston Harbor. A new coal tower (left) was erected in May, 1904. The iron-workers traditionally installed a flag at the completion of their work.

Figure 1-5 – The coal conveyor entered the building, transporting fuel to the coal pockets. Coal then passed through hoppers to the lower level into small coal carts. Coal was then taken to the firing aisles by pushcarts. Photographed July 24, 1903.

Figure 1-6 – The coal handling plant and coal storage pile as they appeared in the 1950s. A large modern steam collier is moored alongside, unloading coal. The thin white structures visible on rooftops, are the enclosures for coal conveyors. (Boston Edison Company Collection, Baker Library, Harvard Business School).

Figure 1-7 – L Street 1898 boilers were B&W Hawley type down draft furnaces, very similar to the ones used by Boston Elevated Railroad Company at their Harvard Power Station. Coal was manually shoveled on flat grates.

Figure 1-8 – A rare photograph dated 1899 showing Boston Electric Light Company's engine room. Note the switchboards on the right.

Figure 1-9 – The engine generator room was finished with white plastered ceilings, white tile floors and walls, while the equipment was painted maroon. Venetian glazed windows diffused the sunlight coming in during the day. Enclosed arc lights mounted on the walls illuminated the room at night.

Figure 1-10 – One of six McIntosh & Seymour vertical cross compound reciprocating steam engines. Here you see the main steam piped into the high pressure cylinder on the left. The low pressure cylinder is on the right side of the machine. The small holes around the periphery of the flywheel were used by the operator for "barring," or advancing the machine off dead center for startup.

Figure 1-11 – One of the many motor-generator sets used for series arc lighting circuits. A steam engine is visible in the background. (Courtesy of Schenectady Museum GE Collection).

Figure 1-12 – Switchboards were designed by Leonard L Elden, electrical engineer for Boston Electric Light Company. The design incorporated some interesting features, such as, the use of aluminum bus bars instead of copper bars. Sections were made of blue Vermont marble mounted on wooden frames, instead of the more common iron frames in use at the time. Wooden parts were polished and stained. Live parts in the back of the switchboard were separated from the wooden frame using thin sheets of asbestos. The bottom switchboard was used for plugging in arc lighting circuits in series. Empty holes are all that remains. (Savaria 2006).

Figure 1-13 – Close-up of the Elden patented automatic (current overload) circuit breaker. The current sensing coil at the base of the breaker tripped the breaker on overload. Breakers were manufactured by Sears B. Condit of Somerville, Mass. (Savaria 2006).

Figure 1-14 – One of the original arc light fixtures retrofitted to accommodate incandescent lamps. Notice the reddish ceiling tiles overhead: the white plaster was removed when it began peeling off. (Savaria 2006).

Part 2 | Historical development of the steam power generating system associated with chimneys C2, C4, and C6.

Figure 2-0 – A classic photograph of L Street Station taken after 1908 with three chimneys C2, C4 and C6. The second coal unloading tower is barely visible hiding behind the original tower. (Boston Edison Company Collection, Baker Library, Harvard Business School).

System Description

Figure 2-1 – Close-up of L Street Station, 1904. (Boston Edison Company Collection, Baker Library, Harvard Business School).

Edison Electric Illuminating Company of Boston (EEI) had initially planned to build a new central AC generating station, Station 4, with an ultimate capacity of 60,000 kw, alongside the existing station discussed in Part 1. The first phase of development consisted of generating 10,000 kw of alternating current, by installing two revolutionary Curtis vertical steam turbines. Construction began in 1903.

The length of the turbine room was 650 feet long, 68 feet wide, and 56½ feet high. There were no basement and no side windows, with daylight being provided by a skylight, known as a "monitor." Large end windows provided good ventilation. The length of the boiler house was 640 feet long, 149½ feet wide and had the same height as the turbine room. The arrangement of boilers was intended to divide the building into seven fire rooms. Skylights were installed in each spacious and well ventilated boiler room. The first switchhouse was 605 feet long, 30 feet wide, and several stories high. The end of the building facing the water was the front of the station. The main entrance into the plant was through offices between the old and newer buildings.

Special attention was given to the arrangement or grouping of apparatus. The turbine room would house all equipment in the plant, including feed pumps, usually considered part of the boilers. Boilers, fuel, and piping were located in the boiler house.

Unlike other Edison plants in Chicago (Fisk Street) and New York (Waterside), electrical equipment at L Street was installed in a separate building alongside the turbine building. This arrangement isolated electrical apparatus from noise and dirt. It also isolated the equipment from damages caused by fires or accidents. EEI wanted the operating room staff—the control operators—to be isolated from the noise and distraction of machinery in the plant. If need be, the operator could always step through doors in the side walls of the room, onto observation galleries. Operators were expected to control generating equipment remotely, with little or no visual contact with turbine generators.

The interior of the turbine room was magnificent. It was finished with tile and enameled brick. The floor was tiled in dark red with black borders arranged to break down monotony of one color. Walls had a wainscot of two-colored green tile about ten feet high. Further up, walls were paneled with a light-colored tile and enameled brick. The crane track was concealed by the wall finish. The entire turbine building was divided into three parts, each separated by a permanent wall and a large door. These doors were usually left opened but were intended to close in case of a serious accident or steam leak in the adjacent room. This had the advantage of keeping construction noise and debris away from operations during plant expansions. Credit for the layout of the station in this fashion is attributed to the company's celebrated president, Charles L. Edgar.

Boilers were arranged in one row directly behind the turbine. Each turbine had its own steam supply but steam lines were connected together to allow cross-connections between steam equipment.

The first two turbines were GE Curtis vertical steam units rated at 5,000 kw. They were four-stage machines with a surface condenser built on the base. Turbine #1 had 15 sets of nozzles on each side. The discs were 13 feet in diameter and revolved at 514 rpm. A ten-inch pipe delivered steam to the turbine on each side of the nozzle chest.

Worthington condensers were designed to use 70 pounds of condensing water per pound of steam, and guaranteed to condense 150,000 pounds of steam per hour while maintaining a 28-inch vacuum with the cooling water at 70 °F.

The original generators were three phase and had Y connected windings. The 60 cycle machines generated 6,900 volts.

Auxiliaries for each turbine consisted of a circulating pump, a wet and dry vacuum pump, a step pump, a hydraulic accumulator, and the feed pump associated with the boilers. All these machines were steam-driven with the exception of the vacuum pump, which was motor-driven.

The weight of the rotating parts of the turbine and generator was about 70 tons. Riding on a film of water, this weight turned with so little friction that it revolved for nearly four hours after the steam was shut off. A unique feature at L Street, the first of its kind, was the mechanical brakes used to bring the unit to rest quickly. With this brake, a machine could be brought to rest in five minutes. The lower step bearing of the turbine was lubricated by water instead of oil. Water was forced into the step bearing under pressure of 1,000 psi by a steam-driven pump. A motor-driven step pump was provided as backup, starting automatically if the steam driven pump failed. To prevent fluctuation in pressure by action of the pump, a surge tank or "accumulator" was connected to the piping system. A heavy weight attached to the accumulator would gradually drop down in cases where the pump failed. The accumulator, when fully raised, had enough water to last the step bearing for ten minutes.

Cooling water pumps took suction from water tunnels running down the center of the turbine room. Two tunnels were provided, each supplying one half of the station. Water intake racks and screens kept out the debris. A wing dam near the shoreline prevented the warm discharge water from mingling with the cold intake water.

A four-inch motor-driven volute pump, running at 1,200 rpm, removed the condensed steam. This pump delivered the condensed water to a National feed water heater. Water then flowed to the feed pumps then into the boilers. Each turbine had a cylindrical hot well and water heater mounted vertically along the wall of the turbine room. Boiler feed pumps were located underneath the heater.

Rows of boilers with superheaters, manufactured by B&W, were placed in pairs with a chimney for each group. The original design called for six chimneys, each 230 feet high, as shown in Figure 1-1. Boilers were elevated from the ground to provide sufficient space for removal of ash. There were no economizers.

Electrical System

EEI initially planned for two bus voltages, 2,300 and 6,900 volt. The company later adopted 13,800 volt as standard. Copper bus bars for each voltage were installed in duplicate or redundant. Each bus was divided or sectionalized by switches. Each section of bus connected to four generators that were connected to the bus bars through one main switch, an oil type circuit breaker, and two selector switches. Switches were housed on the third floor of the switchhouse. The original cells for oil switches, bus chambers and instrument transformer chambers were fabricated on site, made of light yellow brick with a fine cement joint. Bus bars chambers were fully enclosed except for small doors in the walls left opened to access bus terminals. Cells for instrument transformers were left opened except for a fireproof doorway for access to components. Generator leads and transmission lines entered the basement of the switchhouse in ductbanks. Cables were pulled then terminated underneath their respective selector switches. Lead covered paper insulated cable was company standard.

The operating room was on the top floor of the switchhouse and it housed all electrical instruments, meters, protective relays and controls for generators and transmission lines. Power switches and circuit breakers were remote controlled, while generator voltage and excitation were manually adjusted. DC power for excitation was distributed from switchboards mounted along the walls of the operating room. Originally, the 125 volt DC supply for excitation and breaker controls was furnished by steam-driven generators, that were later replaced with motor-generator sets. L Street also had the 125 volt storage battery, mentioned in Part 1, floating on the excitation bus.

Switchboards in the operating room were made of blue Vermont marble mounted on pipe frames. Panel sections were 1½ inch thick, with a bevel on all front edges.

The power distribution system was later redesigned to accommodate separation and segregation of duplicate busses. This involved rebuilding the entire 13,800 volt system in separate fireproof switchhouses. In 1937, the 13,800 volt switching system had to be redesigned to withstand higher short circuit currents. Sixty three breakers were replaced and 46 new concrete cells were constructed.

The Operating Organization

The following describes the day-to-day activities and duties of people working at L Street. The information was extracted from The Electrical World, May 27, 1909.

The L Street Station was divided into an electrical organization and a steam organization. The work of the electrical division was to maintain constant voltages at the feeder terminals throughout the entire system, and to load, operate, and maintain all electric equipment properly. Minor repairs were done in-house; outside contractors were hired for major work. The chief operator was responsible for operating all electric systems, which included the reliable and economic generation of electricity. He worked closely with the chief engineer of the steam division to meet changing conditions.

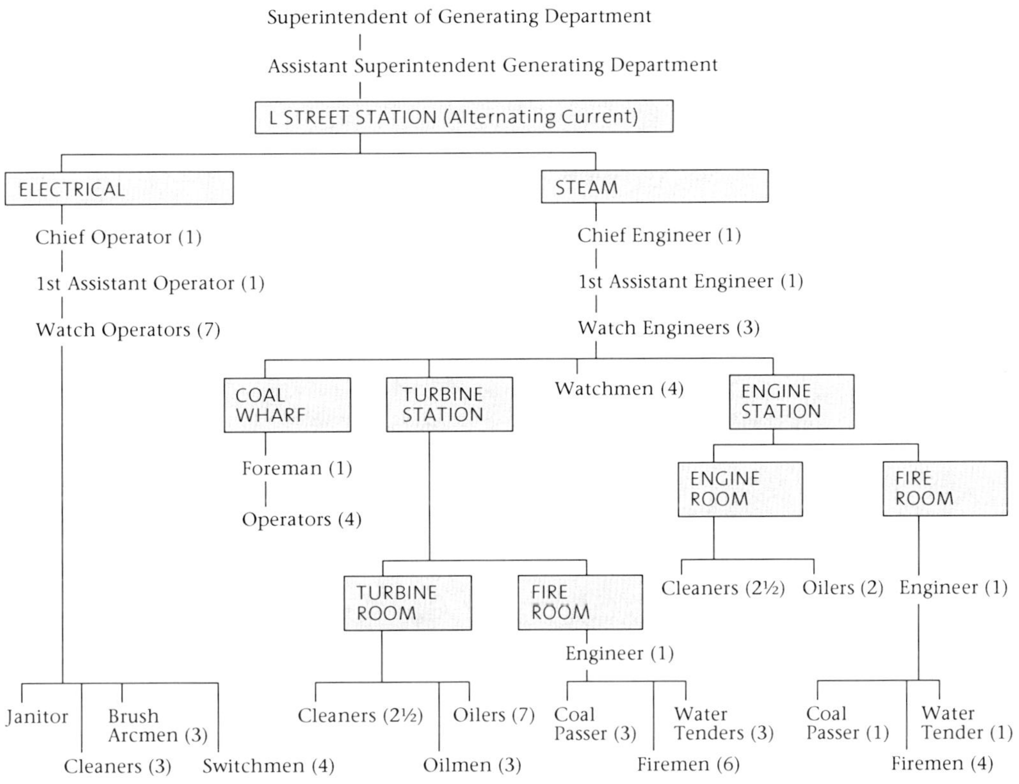

Figure 2-2 – 1909 organization chart for EEI's station staff. Derived from Electrical World, May 27, 1909.

The watch operator on each shift was in charge of the electrical systems in both the engine house and turbine rooms. He was assisted by switchboard men and cleaners. On a given shift, one operator working at the switchboard handled the turbine plant. As shown on the chart, switchboard and Brush arc men were assigned to the engine station. The cleaning force did its work throughout the facility as directed by the watch operator. Two men took care of the forty arc generators, twenty synchronous motors, and three exciters in the engine room. A third man assisted in adjusting and regulating voltages on the 2,300 volt distribution. Arc sets were inspected and cleaned daily using 75 pounds of air to blow out the apparatus.

The chief engineer was responsible to the head of the generating department for the satisfactory and efficient operation of the steam plant. He was a first-class mechanic with wide experience in steam power plants. The chief engineer had similar responsibilities as that of the electrical operating chief, the two working closely at all times.

Daily activities centered around:

Faulty engine cranks or wrists, sticking cylinders, adjustment of pistons, packing leaky valve or glands, maintaining steam piping, supervising the water-line operation, the working of air and circulating pumps, jet and surface condensers, excessive consumption of cylinder and engine oil, the operation of the oiling and step-bearing systems, dry-air and wet-vacuum pumps, number of boilers in service, and quality of coal in use.

Oilers took care of lubricating the machinery and would alert the watch engineer of any unusual condition. He would wipe down and clean the equipment when it was shutdown. One oiler took care of one turbine. The cleaners made the plant safe, attractive and pleasing to the eye.

Each boiler room had an engineer in charge, who was a first-class licensed stationery engineer. He kept in touch with the watch engineer for changing steam demands. The boiler room engineer initiated needed repairs in the fireroom, such as, boilers, headers, valves, furnace brickwork, and grates. Water tenders were in charge of the fireroom watches whenever the boiler room engineers were off duty. Each had a number of firemen assigned to them. Among the principle duties of the water tenders were monitoring and regulating boiler feed water. They took steps to meet increasing or decreasing steam demands by adjusting water flow. Their duties included scheduling when to bank, fire, light or clean boilers. They reported steam or water leaks to the watch engineer. The firemen were described as, "selected men of reliable temperament," whose primary duty was to burn coal carefully. This involved judgment in both firing and cleaning. The firemen was expected to know the character of the coal coming into the boiler room, then regulate the depth of stoker feed and adjust the speed of the grate, depending on steam demands. The coal-passers handled the fuel and cleaned the boiler room.

The chief electrical operator and the chief engineer of the steam division each had a clerk to assist in record keeping. Different parts of the station were interconnected by telephones. EEI had contracted with New England Telephone Company for a PBX (private branch exchange) to be centrally located at L Street. Telephones were connected to all substations and the general offices of the company. The PBX was manned 24/7 by three telephone operators.

Personnel in the operating room and the turbine room coordinated their actions using a mechanical Cory engine-telegraph system, like those aboard ships. The operator in the operating room transmitted signals to the turbine floor (ready, start, stop, turn, speed, trouble, or standby). The watch engineer executed the action then acknowledged the signal. A telephone station was also available for the operator on duty. As backup communications, the operating room was equipped with a megaphone.

The operator in the operating room controlled each turbine-generator from an individual control panel. Excitation was manually adjusted. In the turbine room, the watch engineer took charge of the throttle valve by standing at the turbine, manually starting and stopping the machine based on signals received. Each telegraph station was floor mounted on a pedestal. Multiple telegraph stations were located in the operating room. Each turbine room had its own telegraph station.

Accidents

Making electricity has never been easy and it was especially hazardous in the early days. The following accidents were reported by the Generating Department for fiscal year 1906-1907.

1906:

- — August 5: Chas. C. Buchanan, oiler; jammed fingers in the engine gear. Ends of two fingers amputated. Was away from work 23 days.
- — August 6: Mr. Harahan, coal trimmer; overcome by heat. No time lost.
- — August 13: W. A. Robertson, boiler repairer; slight burn on arm. Was out four days.
- — September 27: Fletcher Oakes, oiler; fingernail torn off in engine gear. Time lost: one hour.
- — October 2: M. L. Fell, oiler; thumb cut in pump. No time lost.
- — October 4: Albert Babcock employed by Lundin Electrical & Machinery Company working on coal belt. Bruised by pulley falling on it.
- — October 18: Albert Pfefferle, battery man; arm and hand burned by short circuit. Was out ten days.
- — November 25: L. H. Howe Jr., switchboard man; hand bruised between switch and wall. No time lost.

1907:

- — January 4: Charles A. Patterson, switchboard man; flash burn from a dropping cummutator. Was out 2 days.
- — January 8: Jos. P. King, assistant chief operator; burned and bruised by contact with fuse clip. Was out 33 days.
- — January 12: S. W. Choate, oiler; finger broken in engine valve gear. Time lost: part of a day.
- — February 7: Thos. O'Meara, teamster employed by O'Conner Brothers. while moving machinery. Finger jammed.
- — February 13: Jas. W. Emery, oiler; finger jammed in engine valve gear. Was out 23 days.
- — February 18: Charles S. Hadaway, chief operator; hand scorched while pulling switch. No time lost.
- — March 4: CB Wilton, water tender; face hurt by bursting water gage glass. Was out three days.
- — April 22: John Pierce, fireman; fell and injured knee. Was out 15 days.
- — May 18: P. Morgan, oil man; fell and bruised face and groin. No time lost.
- — June 1: I. A. Tenger, watch engineer; hand cut by grinding machine. No time lost.
- — June 10: L. M. Haggerty, switches man; hand burnt by shock from switchboard. No time lost.

Startup History

Station operating personnel kept detailed equipment journals during final construction and the commissioning phase. It may be interesting to include a few journal entries as part of L Street's history. The following entries have been taken at random from journals archived in the Boston Edison Collection, at Harvard's Baker Library:

Starting Turbine #2:

- — October 21, 1904: first started. Up to 450 rpm at 1:30 p.m. Balancing poor and stopped at 4 p.m.
- — October 31, 1904: 9 a.m. with 1,000 kw. Rebalanced. Governor problems.
- — November 3, 1904: General Electric Company (GE) sent governor to Schenectady for factory adjustments.
- — November 7, 1904: reinstalled governor and running 1,000 kw. Valve sticking.
- — November 8, 1904: 3,600 kw. Regulating well.
- — November 12-18, 1904: Mr. Krusie from GE is here to examine step bearing. Turning down 2nd & 3rd stage wheel. Working on bearings, buckets, etc.
- — November 23, 1904: restarted turbine #2 at 10:30 a.m. with load at 11:20 a.m. up to 6,000 kw. Generator went badly out of balance and causing rubbing at the above load. Shutdown 4:45 p.m.
- — November 25, 1904: GE balancing unit 2. Top bearing down.
- — November 27, 1904: emergency tested 550 rpm. 4,000 kw; running good. Water rheostat ready to synchronize with old station.
- — November 29, 1904: first load on #2. 6,000 kw. Turbine #1 online at 1 p.m.; #2 put in at 4:15 p.m. Operating both in parallel. Stopped #1 at 6 p.m. Turbine #2 finished the run with 4,600 kw.

1912 Inventory of Station Equipment

In 1912, the inventory of power equipment in the steam turbine division included:

Boilers:

- — Thirty six 512 hp B&W boilers
- — Eight 508 hp B&W boilers
- — Total boiler capabilities: 22,496 hp

Grates:

- — Twenty two Roney stokers, 22 Murphy stokers

Stream Throttle Pressure:

- — 200 psi at 380 °F.

Prime Movers:

— Three 12,000 kw, 6,900 volt, 60 cycle, 3-phase, 720 rpm, Curtis vertical turbo-generators

— Two 15,000 kw, 6,900 volt, 60 cycle, 3-phase, 720 rpm, Curtis vertical turbo-generators.

The potential of the last two units was raised to 13,800 volts by using transformers.

Auxiliary Apparatus:

— Two 300 kw 60 cycle synchronous motor generator exciters, 6,900 volts AC to 125 volts DC.

— Plus a variety of single-phase transformers.

Plant Additions and Modifications

L Street Station was continuously upgraded and underwent many different kinds of modifications over the years. Here is a list of the most noteworthy projects:

1905:

— Installed turbine T #3.

1907:

— Installed turbine T #5.

1908:

— Added a new steam operated coal tower.

1909:

— Two new turbines rated 14,000 kw replaced the original 5,000 kw units. Converted the 4,600 volt switchboard to 6,900 volt. New lines and a 6,000 kw step up transformer were installed, including a 5,000 kw transformer, 14 GE motor-operated switches and motor generator sets, and one 300 kw motor generator exciter set for T #1 and #2.

1911-1912:

— New turbine T #7, rated 15,000 kw was added to the station. This required a new steam supply, complete with steam and water piping, valves, and foundations.

1912:

— A signaling system between the boiler rooms and the operating room was added. The work consisted of illuminated signs, 31 key taps, 200 four cp (candle-power) lamps, a Mason monogram carriage call, switches and a clock.

— A change room was added on the third floor of the station. Other facilities were added for the Load and Dispatchers Department, including sleeping quarters for dispatchers.

— An additional telephone set was added between the operating room and the weighing tanks in the basement of boiler room.

1915:

— A Gamewell fire alarm system was installed.

1916:

— A "wireless telephone and telegraph outfit" was added to the station.

1916:

— Westinghouse stokers were added under boilers #13 to #20, replacing the original stokers. The armature on generator T #6 was rewound for 14,000 volts.

1917:

— The guard house was erected, complete with furniture, light and speaking tubes.

1921:

— Sixteen Riley underfeed stokers were installed for boilers #37 through #52. New soot blowers and Bailey meters were also installed.

1922:

— Each of the eight vertical turbines was retrofitted with over speed tripping devices. A major modification project was also undertaken which involved raising the ceiling and monitor in Switchhouse #1 by one level. Reactors were first introduced into transmission line circuits.

1926:

— Alarm indicators were provided in each of the three turbine rooms to alert watch engineers that protective relays on turbo-generators had actuated. Alarm circuits were 125 volt DC circuits using illuminated signs and a Klapton horn in each turbine room.

1928:

— A first aid room was built and furnished with sterilizer, hospital bed, hair pillow, rubber blanket, and hair mattress.

1929:

— Generator T #1 was rebuilt, increasing its rating to 12,000 kva and raising voltage to 13,800 volts.

The Warren Clock for Frequency Regulation

Henry Ellis Warren, an inventor and clockmaker from Ashland, Mass., brought Greenwich Mean Time (GMT) to the control rooms of power generating companies. Warren invented the Warren Master Clock. It gave control room operators a way to adjust and regulate frequency based on a known standard. This was accomplished by giving operators a simple benchmark; a point of reference to guide them whenever adjusting generator frequency. The Warren Clock was introduced at L Street in 1916. General Electric Company quickly acquired the invention then began marketing Warren Master Clocks throughout the US power industry.

Before Warren's invention, system frequency would fluctuate widely by 0.5 percent of a cycle or more. Using conventional frequency meters, operators did the best they could by tweaking turbine speed. One day, while working on his electric clock, Warren noticed that Boston Edison's frequency was running slow. His precise time measurements indicated that the AC service was off by one-half cycle, resulting in a loss of ten to fifteen minutes a day. He called L Street Station to complain. EEI engineers were intrigued with Warren's findings and he was invited to come in. Warren's invention used two five-minute clocks: one clock actuated by a small synchronous motor, the other, a precision timepiece set to time in Washington, D.C. By comparing the difference between the two clocks, the operator was able to adjust the speed of the turbine more precisely than ever before. The first Warren Clock was installed and placed in service October 23, 1916, making L Street Station the first to generate perfect 60 cycles.

The December Peak of 1908

The climax of the station year was the December peak. The maximum load on the L Street plant was about 28,000 kw at 5 p.m. December 23, 1908. Both the steam engine and the turbine plants were ready for service. There were 9 engines and turbo-generators in operation at 5 p.m. that day, with a minimum of three online during the day. Each engine generator was capable of 1,500 kw. Turbines #3 and #4 were 5,000 kw, and turbine #5 was rated 8,000 kw. The normal rating of the generating units in service at 5 p.m. was 31,000 kw, with the largest turbine at its guaranteed continuous capacity of 12,000 kw. Ninety one men were on duty throughout the peak period.

Photographs and Illustrations

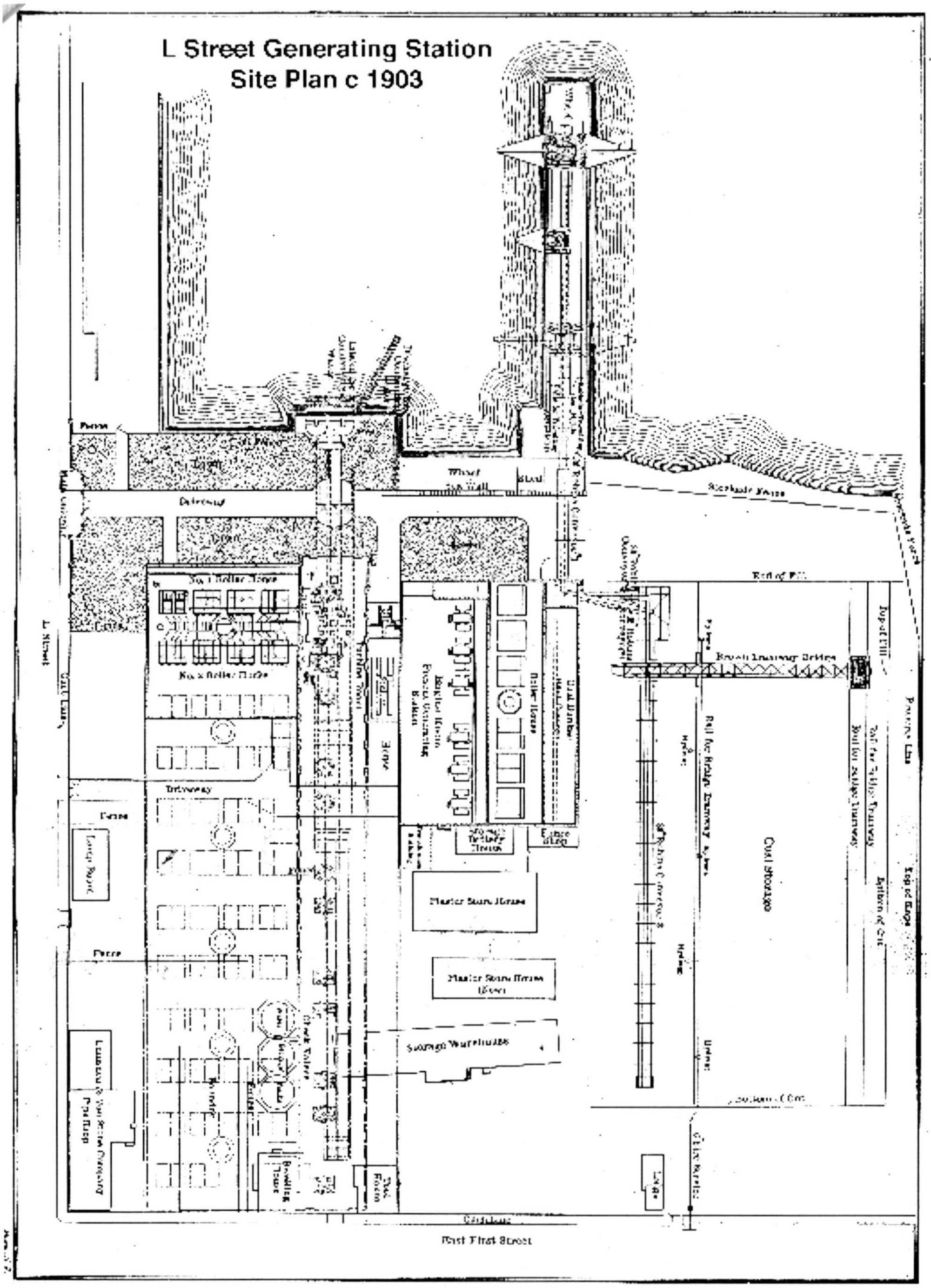

Figure 2-3 – 1903 site plan of the L Street property.

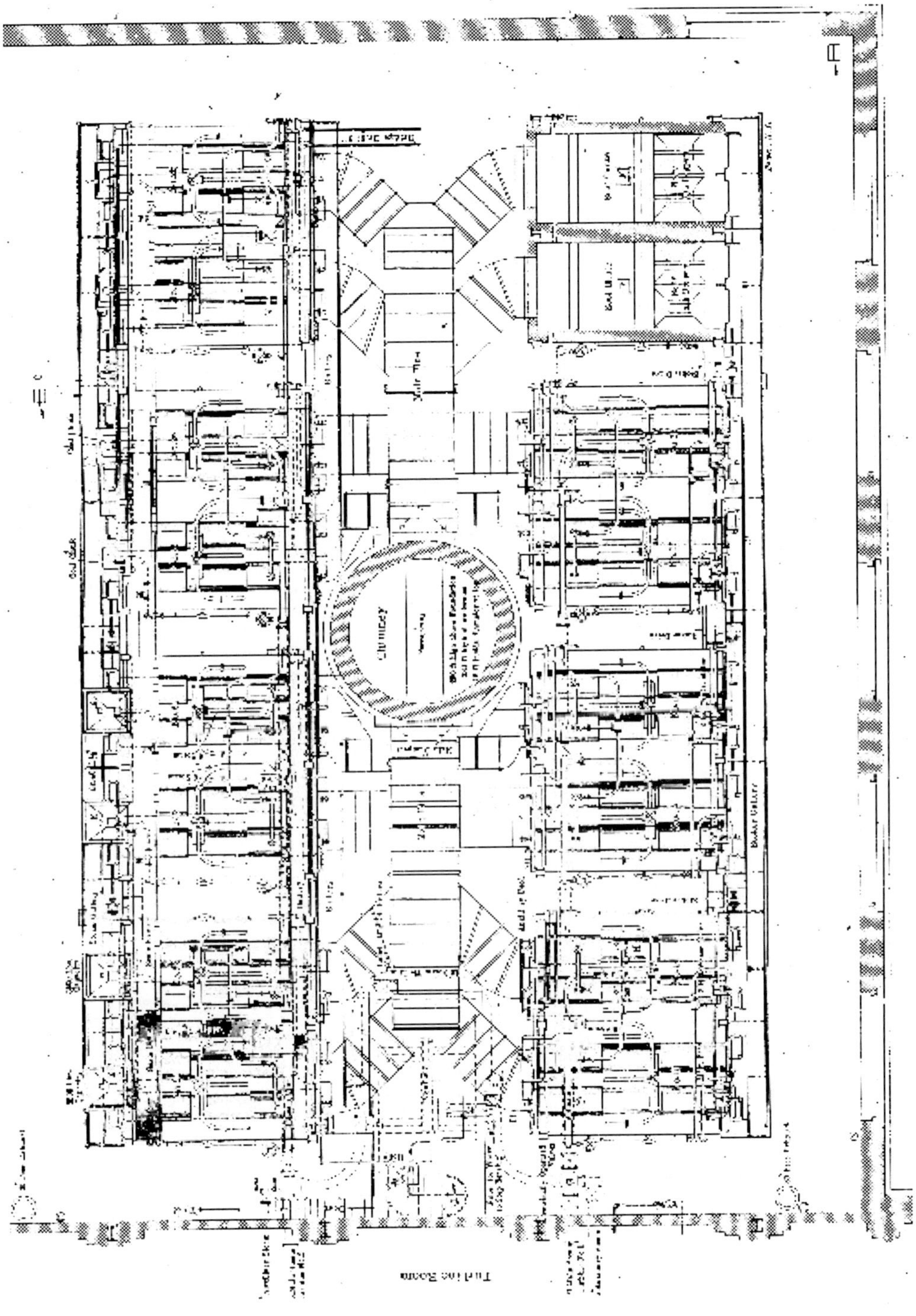

Figure 2-4 – General arrangement of boilers, flues, chimney, steam piping and other auxiliaries.

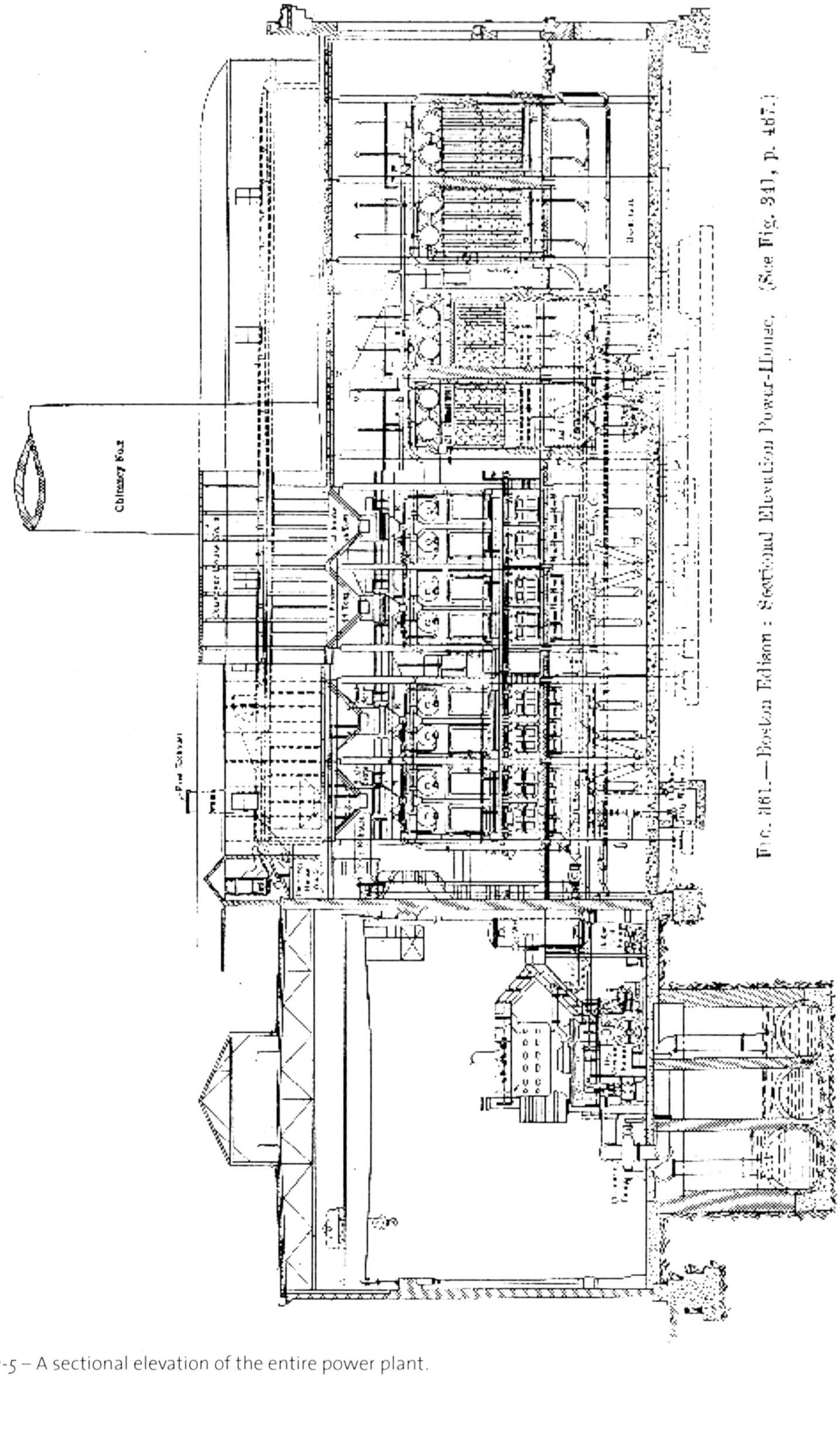

Figure 2-5 – A sectional elevation of the entire power plant.

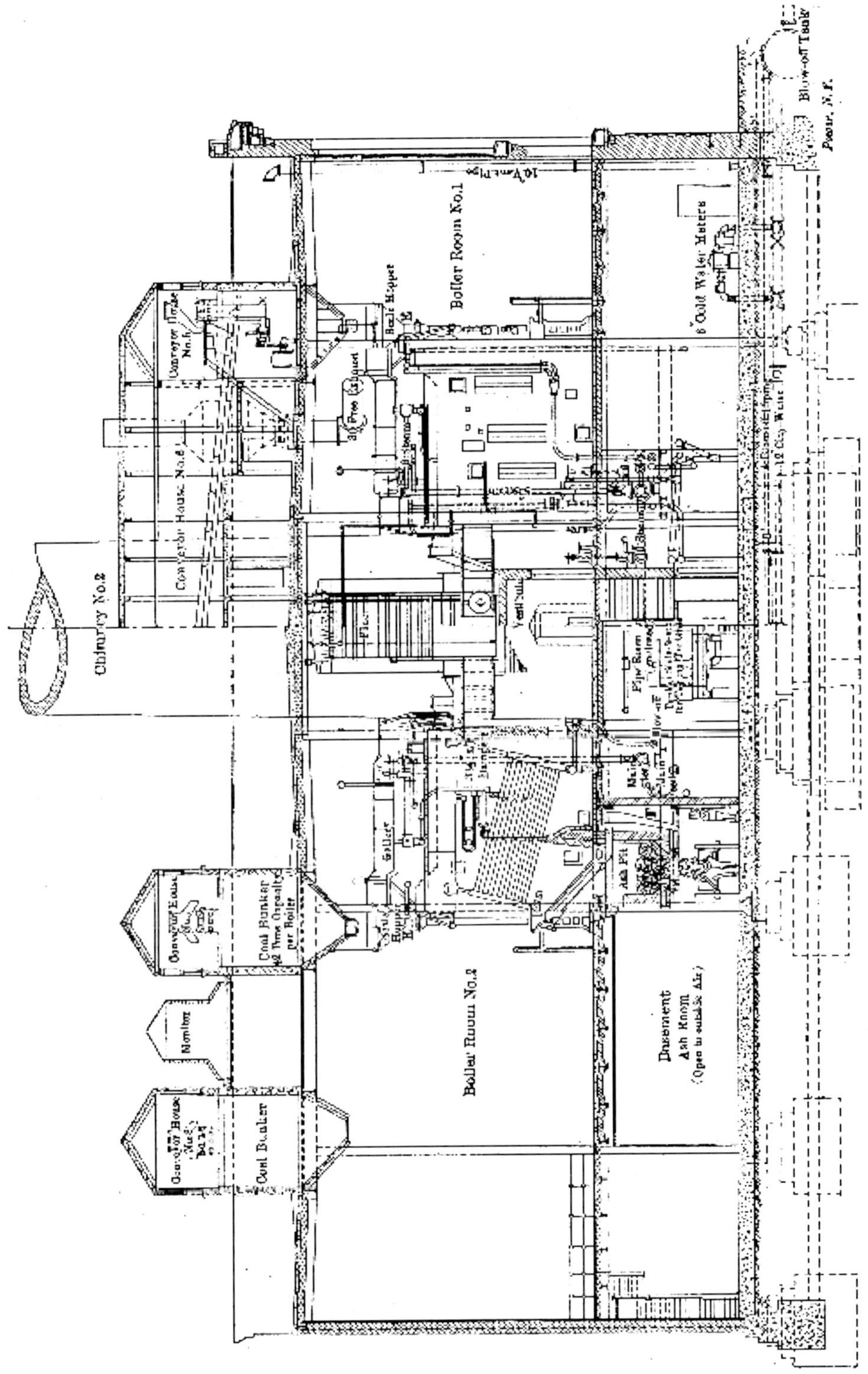

Figure 2-6 – A transverse elevation of the boiler house. Note the horse drawn cart in the basement collecting ashes from the ash pit.

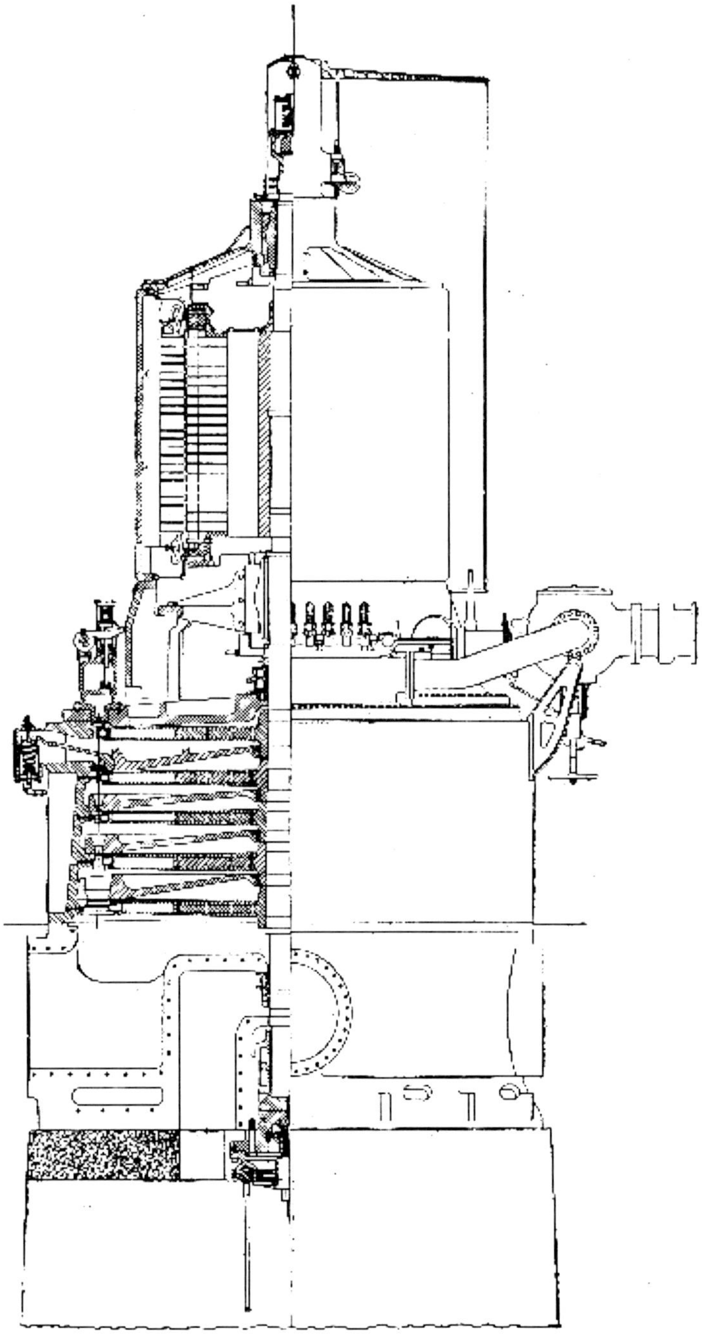

Figure 2-7 – Cross section of a 9,000 kw Curtis vertical steam turbine generator. Note the governor speed changer connected through linkages to the steam admission valves. From a General Electric catalog.

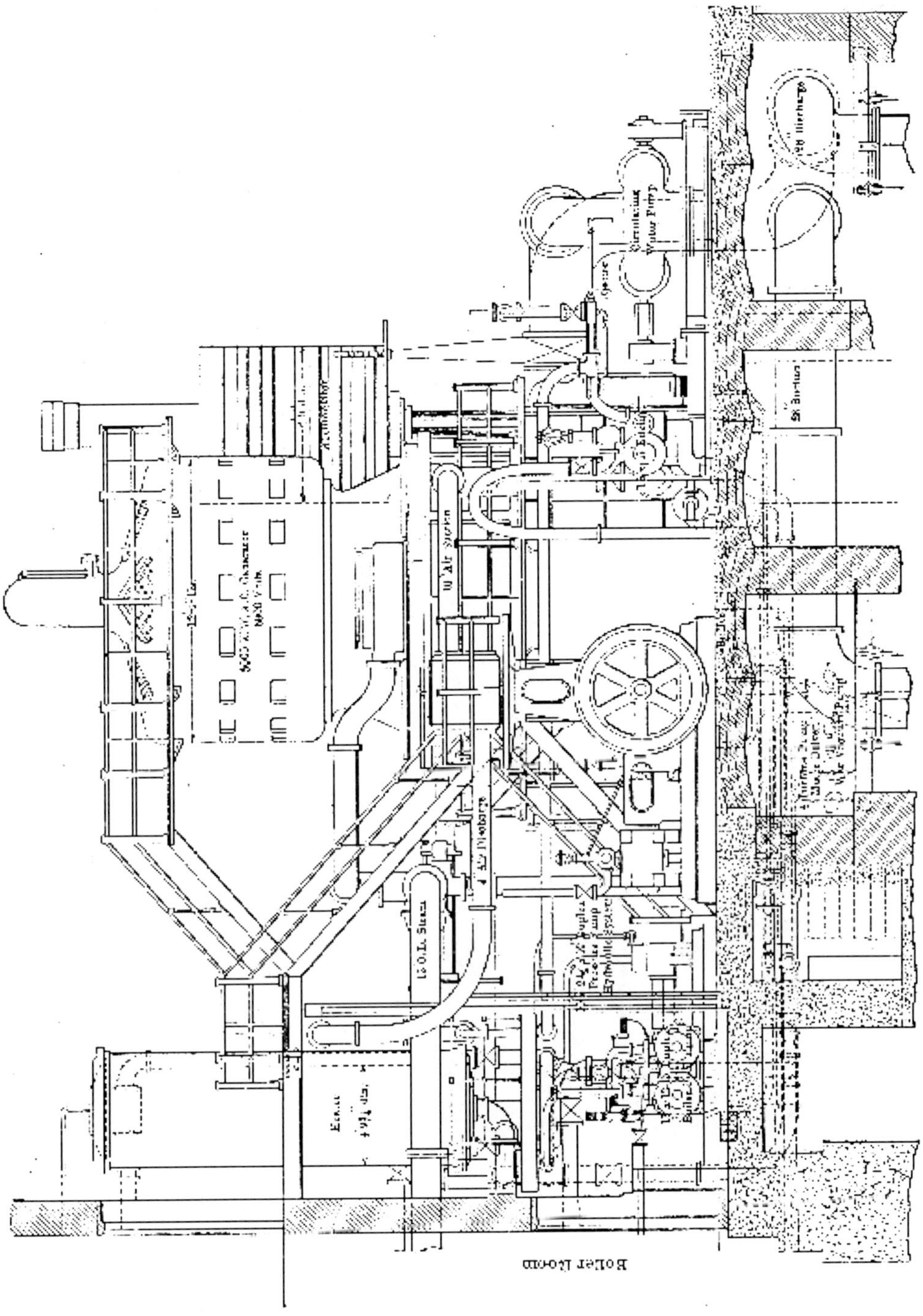

Figure 2-8 – A section through the first 5,000 kw turbine generator installation at L Street, complete with all auxiliaries. The second unit erected was identical. From Power July 1905.

Figure 2-9 – Interior of the boiler room shows spacious firing aisle and space for future boilers.

Figure 2-10 – A battery of boilers in Boiler Room 1. Roney stokers were mounted high off the floor, requiring a platform along the front of the boiler. The motor drives for the stokers are not very clear. Each boiler was equipped with a coal weighing hopper. Coal was dropped down the pipe into the hopper in front of the furnace.

Figure 2-11 – Turbine #1 was photographed many times by equipment manufacturers, engineering organizations, and consultants writing books on the latest in steam turbines. This particular photograph was taken in 1907 by Niles, the crane manufacture, for their trade catalog.

Figure 2-12 – Turbine Room 1, looking north at turbine T #4. Turbine T #1 is farthest away against the window.

Figure 2-13 – Turbine Room 2 with one of General Electric's larger 12,000 kw

Figure 2-14 – Ten members (look carefully) of L Street steam turbine division posing alongside turbine T #3. The picture was probably taken in 1906 or 1907. (Boston Edison Company Collection, Baker Library, Harvard Business School).

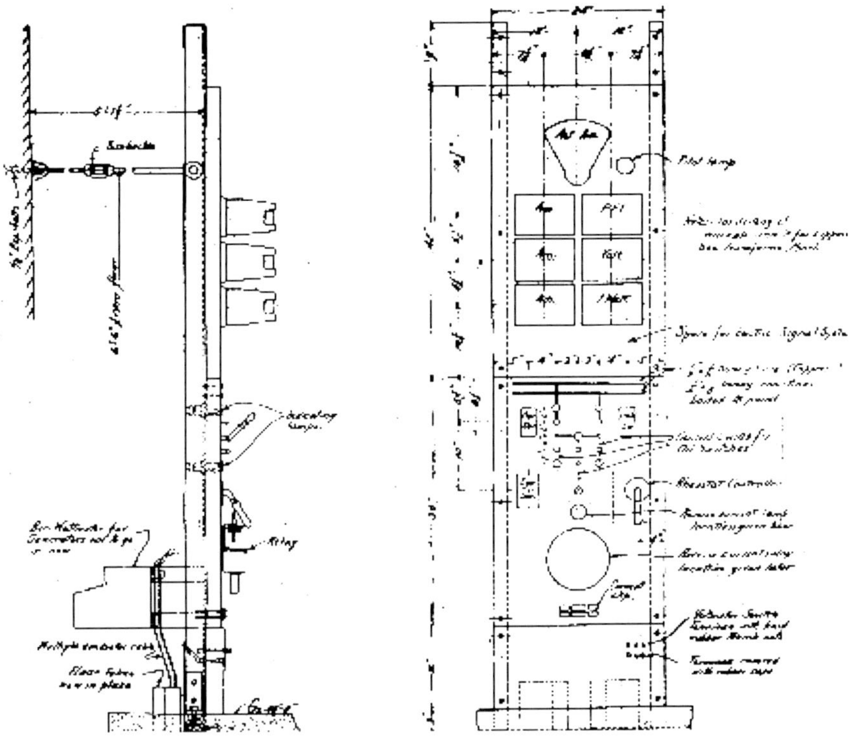

Figure 2-15 – One of the turbine generator control panels in the operating room that occupied the top floor of the switchhouse. These panels were made blue Vermont marble supported on pipe frames. The top of the panel had a six-inch green decorative strip of ceramic tile molding. The green molding matched the tiles in the turbine rooms.

Figure 2-16 – A rare view of the operating room showing the control operator at his signals. The operator stayed in close contact with the watch engineer on the turbine floor during startup and shutdown maneuvers. On the left side of this photograph, there are five turbine generator control panels. The switchboards behind the operator are part of the 125 volt DC supply for excitation and breaker controls.

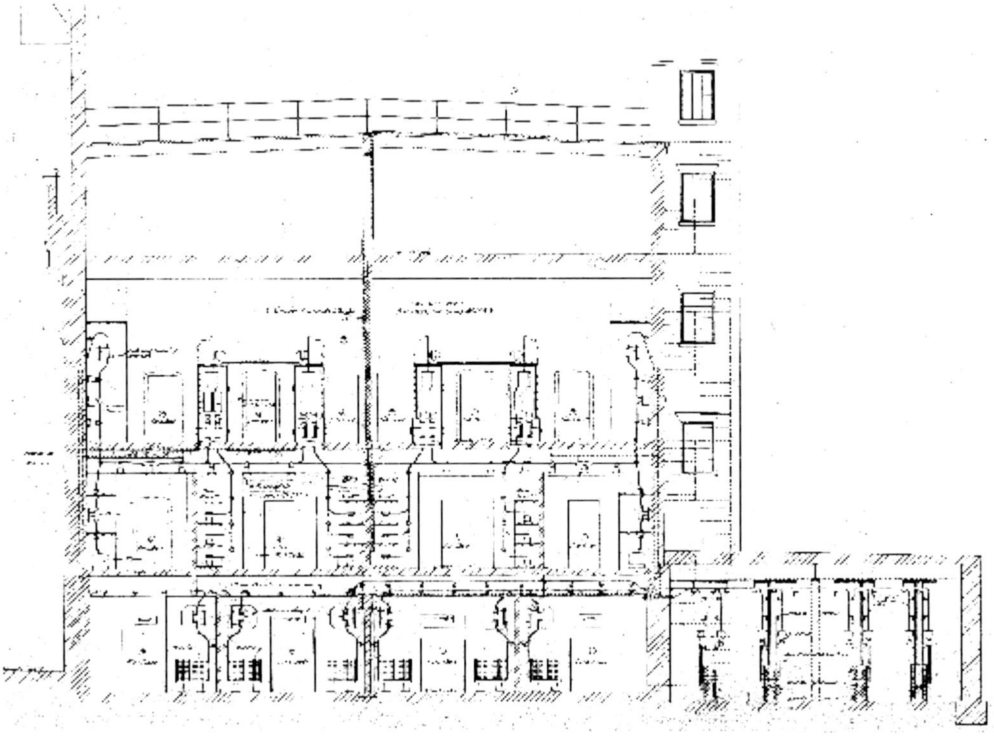

Figure 2-17 – Cross section of the switchhouse shows how the 6,900 volt and 13,800 volt switching systems were built before the advent of pre-fabricated metalclad switchgear. Components, switches, circuit breakers, bussing, and power cables were arranged vertically in multi-story buildings. The electrical site contractor assembled busses and components in concrete cells.

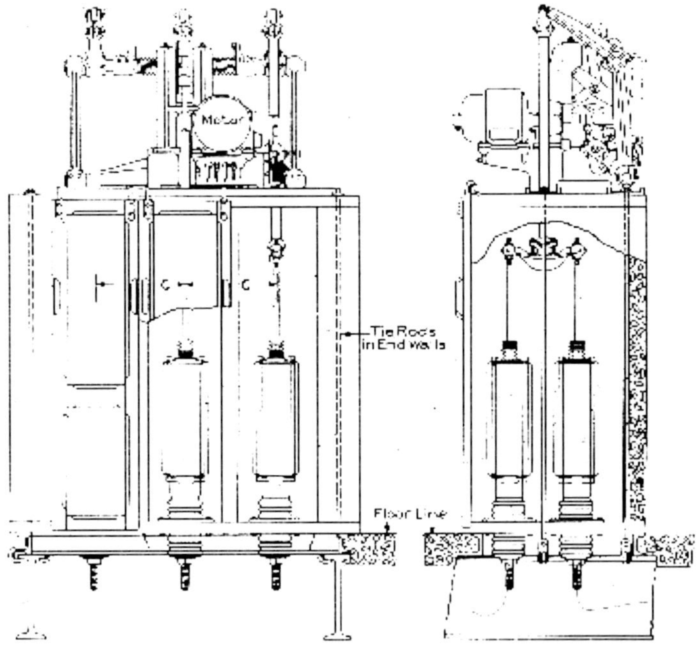

Figure 2-18 – GE oil circuit breaker type H-6. Breakers were tripped opened by solenoid, and closed by a DC motor.

Figure 2-19 – Control panel for turbine generator T #6. Each breaker was remotely controlled using knife switches to complete the 125 volt DC open-close circuits. Note the newer gray panel on the right made of metal plate.

Figure 2-20 – A passageway through one of the switchhouses.

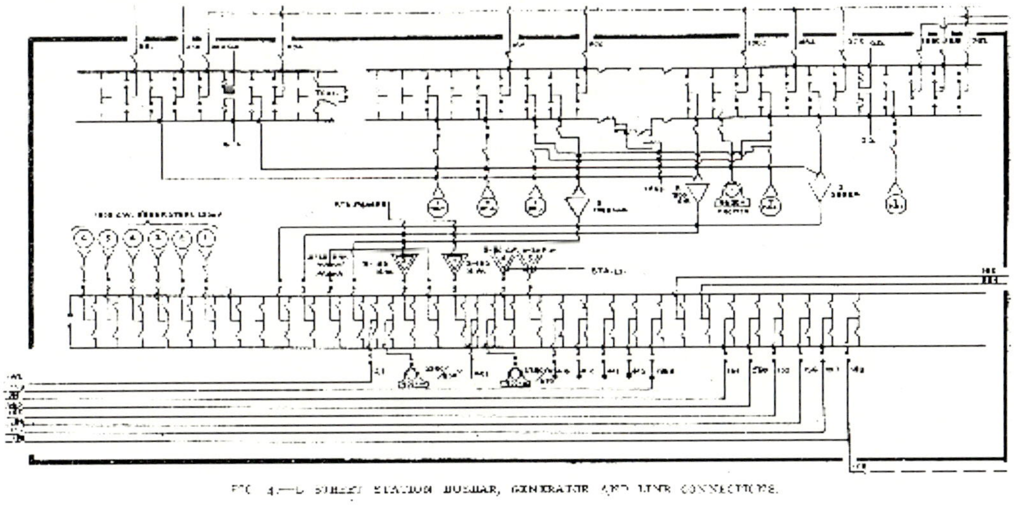

Figure 2-21 – 1909 AC electrical one-line diagram.

Figure 2-22 – The keystone above the great arched window bears the figure of Mercury. Just above it is the word Edison and Roman numeral MCMIII (1903). (Savaria 2006).

Figure 2-23 – A sample of the interior, designed by Winslow and Wetherell, architects of Boston (Savaria 2006).

Figure 2-24 – The north window from inside the turbine room. (Savaria 2006).

Figure 2-25 – A 120–220 volt AC fused distribution panel. Copper bus bars are secured to a slate backboard. Panel trim and door are made of wood. The door is fireproofed with sheets of asbestos. (Savaria 2006).

Figure 2-26 – Lineup of L Street's early turbine-generator control panels. (Savaria 2006).

Figure 2-27 – The steam-driven vacuum pump for Turbine #8. (Savaria 2006).

Figure 2-28 – Personnel from New Boston Generating Station posing with Turbine #8, the last surviving vertical turbo-generator. (Savaria 2006).

Construction Photographs – Period 1903 to 1912

Excavation work with Boston Electric Light Company's power station in the background. March 17, 1903

Installing fence along L Street. September 23, 1908

Installing tubes in turbine #5 condenser. July 29, 1908.

Excavating by hand. May 21, 1903.

Building the first chimney (chimney #2). Bricks were molded to conform to inside and outside chimney diameters. May 27, 1903

Chimney #2. July 3, 1903.

Chimney #2. July 16, 1903.

The power station grew southward towards East First Street. In September 1907, bricklayers were closing up Turbine Room 2.

Looking at Boston Harbor and Summer Street from the top of chimney #2. September 6, 1903.

Looking westward with the 1898 chimney in the foreground. September 9, 1903.

Erecting steel for the boiler house. August 3, 1903.

Boiler house. August 26, 1903

North side facade going up. November 2, 1903.

The boiler house seen from L Street. October 24, 1903.

Duct banks and manhole. December 16, 1903.

Progress on the electrical room. February 11, 1904.

Interior of the turbine room. February 20, 1904.

Facade with lighting fixtures. June 9, 1904.

Driveway for heavy teaming. March 30, 1904.

View in electrical cells in the switch room. June 20, 1904.

Moving part of the condenser. June 14, 1904.

Unknown millwright standing inside the armature. July 11, 1904.

View of turbine room. May 25, 1904.

General view of turbine. July 27, 1904.

Curtis vertical steam turbine #1. September 2, 1904.

Electric hoist and boilers. October 22, 1908.

Cory signals for turbines 1 through 4. September 11, 1908.

Electrical operating room. October 16, 1908.

Backside of the power station viewed from the corner of L Street and East First Street. October 1, 1908.

Part 3 | Historical development of the steam power generating system associated with chimneys C8 and C10.

Figure 3-0 – Photograph of L Street Station taken in the thirties with two additional chimneys C8 and C10. A steam collier is discharging coal. Two tall stacks on the left belong to South Boston Power Station, the power plant for Boston's rapid transit system. (Boston Edison Company Collection, Baker Library, Harvard Business School).

System Description

The 1916 L Street Station extension was a major project, which incorporated the latest practice in large power plant design. The boiler room was unequaled in its design, spaciousness, and performance efficiency. The highest commercially feasible steam pressure was utilized. According to power engineers who visited the power plant extension in 1920, the new turbine and boiler rooms were exceptionally roomy, well lit and ventilated. The boiler room housed eight Babcock & Wilcox cross drum water-tube boilers, each with 12,320 square feet of heating surface. Three new horizontal turbines were housed in a new room designated Turbine Room 3. Each turbine had a surface condenser and sat on a concrete pedestal. The three turbine generators were identical, 35,000 kw, single-cylinder impulse design, incorporating General Electric's latest improvements. EEI was one of the first companies to have an exciter directly coupled to the turbine-generator shaft.

B&W boilers, designated #61 through #68, supplied the 300 psi steam system. Each was rated 86,000 lbs/hr. Their flues were connected to chimney C8. In a subsequent phase, four more slightly bigger boilers rated 125,000 lbs/hr were added. These four boilers, designated #69 to #72, shared a new stack, chimney C10.

Boilers were enclosed in individual steel casings with insulated walls instead of bricks. Each boiler was fitted with the latest automatic, 13-retort underfeed stokers with clinker grinders, manufactured by Westinghouse. Retorts were sectionalized in three groups of three retorts, forming a complicated mechanism, best illustrated later in this Part 3. Each stoker was connected to four motors, one for each section of retorts. Motor enclosures available at the time were not dustproof, but were open construction. Boiler feed pumps and force draft fans were all motor-driven and mounted on the boiler room floor.

Each boiler had a steam drum running across the width of the boiler, referred to as a cross drum design, instead of the more common longitudinal design. Boiler drums were riveted construction. In a cross drum boiler, the front of the boiler became the back, as far as observing fires and controlling firing rates. One person took care of feeding coal to all the boilers by operating a motorized 10 ton weighing coal larry. The firemen manned the controls from the side of the boiler away from the stokers. Rheostats, which controlled the speed of the stokers, were manually adjusted by turning a hand wheel from the boiler room floor. Rheostats, motor controllers and other electrical devices were housed in metal enclosures located under the boiler floors, presumably to escape from coal dust. Westinghouse clinker grinders at L Street were of considerable interests to the steam electric industry and were featured in articles and textbooks. Boilers were equipped with individual motor-driven forced and induced draft fans, and came equipped with economizers.

The new steam piping system was designed for a pressure of 300 psi superheated steam. It was connected to the existing 200 psi main header through pressure reducing stations and isolation valves. This piping arrangement is best understood by studying the Steam Flow Diagram available in the next Part 4. To avoid the possibility of 300 psi steam onto lower pressure piping, a manifold of twelve 4½-inch safety relief valves was installed, relieving steam to the main atmospheric relief pipe. L Street had many motor operated valves on steam and other piping systems. These electrically operated valves were noteworthy at the time, based on a novel control circuit designed by PP Dean, and manufactured by Cutler-Hammer Company.

History of the L Street Extension

Modifications, additions and changes were planned and authorized by EEI, then executed by local contractors. Examples of this work are given below.

- — March 31, 1916: $210,000 was authorized to purchase GE turbine T #9.
- — September 20, 1916: A major project was initiated, extending the coal handling and cooling water systems. This included new circulating water tunnels and a screen house. Water tunnels were to extend all the way into a new turbine room.
- — September 30, 1916: Project authorized to complete the installation of boilers, economizers, fans, and steam and exhaust piping, heaters, feed water piping, and miscellaneous. Four B&W boilers were purchased separately.
- — September 30, 1916: A very large building project estimated at $891,000 was approved. This included a new switch house, turbine room, boiler room and a brand new stack, chimney C8.
- — January 9, 1917: A purchase order was awarded to GE to furnish and install a 35,000 kw turbine generator (T #10) for $373,347.
- — August 29, 1917: Appropriation approved for the electrical installation of T #10.
- — October 20, 1920: Purchase order of $510,000 was awarded for a new 35,000 kw turbine-generator T #11.
- —November 3, 1920: A major $1,000,000 project was initiated to extend the turbine room, boiler house, and increasing the capacity of the coal handling system. A new switch house was also included.
- — December 28, 1921: $310,000 was authorized to purchase and install turbine generator T #11. Additional money was authorized for economizer pumps, barometric heater, make-up pumps, main steam and exhaust piping, turbine piping and generator auxiliaries.

Startup History

The following are abbreviated summary notes of activities during the startup of the turbine generators.

Turbine #9 – 1920

- — September 13 & 14: Unit ready for trial balance. Warmed up by midnight. Exciter is on and operating properly and up to voltage. Governor parts arrived from the factory by messenger at 11 p.m..
- — September 15: Unit started at 9:20 a.m. Began to vibrate very badly. Off on emergency, and then back on again. Taking excitation from the bus while the exciter is undergoing tests. Conditions at speed are OK. Placed online for more tests. Balanced perfectly.

— September 19: Running to balance unit with 5,000, 10,000 and 15,000 kw. Operating fair. Shutdown and changed weights.
— September 28: Unit online carrying 1,000 to 2,000 kw.

Turbine #10 – 1921
— March 1: First steam on for tests at 11 a.m.
— March 14: Pulled 26-inch vacuum.
— March 16: Began turning to 500 rpm. Stopped due to hot bearings on air pump.
— March 27: To speed at 8:19 a.m. GE doing heat run.
— April 21: Loaded to 20,500 kw.

Turbine #11 – 1924
— August 9: GE is ready to run turbine #11.
— August 11: Unit online carrying 30,000 kw.

Early Pollution Controls

One of Boston Edison's contributions to power engineering is in the area of pollution control. L Street participated in these advances as far back as 1937. Atomizing water scrubbers were installed in chimney C8, that collected flues from the eight boilers, #61 to #68. The design was based on experiments previously undertaken at the Kneeland Street steam plant in downtown Boston. Chimney C10 was also equipped with a gas scrubber in 1937.

In 1938, Research Cottrell electrostatic precipitators were installed in the stacks at both the Edgar Station and L Street Station. Precipitators were successful in removing over 95 percent of chimney cinders and fly ash. By the 1950s and 1960s, gas scrubber performance was greatly improved by the addition of electronic alarms, television monitors, and radiological monitoring stations.

An editorial in Electrical World September 25, 1937, encouraged electric utilities to follow Boston Edison's lead of freeing flue gases from objectionable particles before discharging them into the atmosphere.

1921 Inventory of Station Equipment and Overall Performance

Boilers:
— Eight 512 hp B&W boilers with Westinghouse stokers
— Thirty two 512 hp B&W boilers with Murphy stokers
— Eight 508 hp B&W boilers with Murphy stokers
— Eight 1,232 hp B&W boilers with Westinghouse stokers
— Total boiler capabilities: 34,400 hp

Turbine Generators:

— Unit 1	720 rpm, vertical, 12,000 kw
— Unit 2	720 rpm, vertical, 12,000 kw
— Unit 3	720 rpm, vertical, 7,500 kw
— Unit 4	720 rpm, vertical, 7,500 kw
— Unit 5	720 rpm, vertical, 12,000 kw
— Unit 6	720 rpm, vertical, 15,000 kw
— Unit 7	720 rpm, vertical, 15,000 kw
— Unit 8	720 rpm, vertical, 7,400 kw
— Unit 9	1,800 rpm, horizontal, 35,000 kw
— Unit 10	1,800 rpm, horizontal, 35,000 kw

1921 Overall Station Performance

— Maximum generating kw at 4:45 p.m. December 15	120,000 kw.
— Average generated kw for the year	39,732 kw
— Load factor (average divided by maximum)	0.331
— Total kw capacity of generators	156,000 kw
— Percent generated to total generated capacity	28.2

Photographs and Illustrations

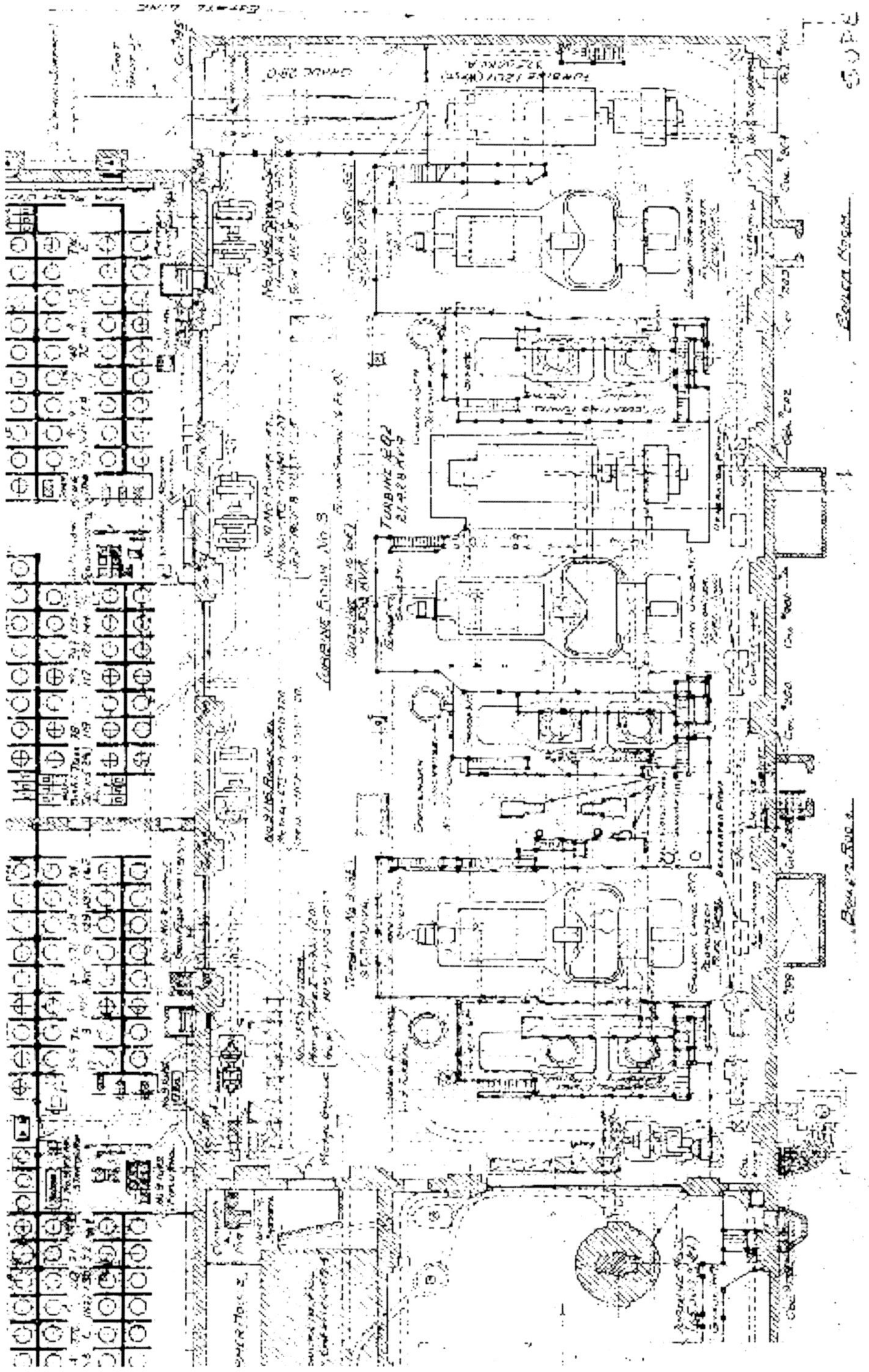

Figure 3-1 – Part of Station General Plan, EEI drawing 12004-Y showing the arrangement of equipment in Turbine Room 3.

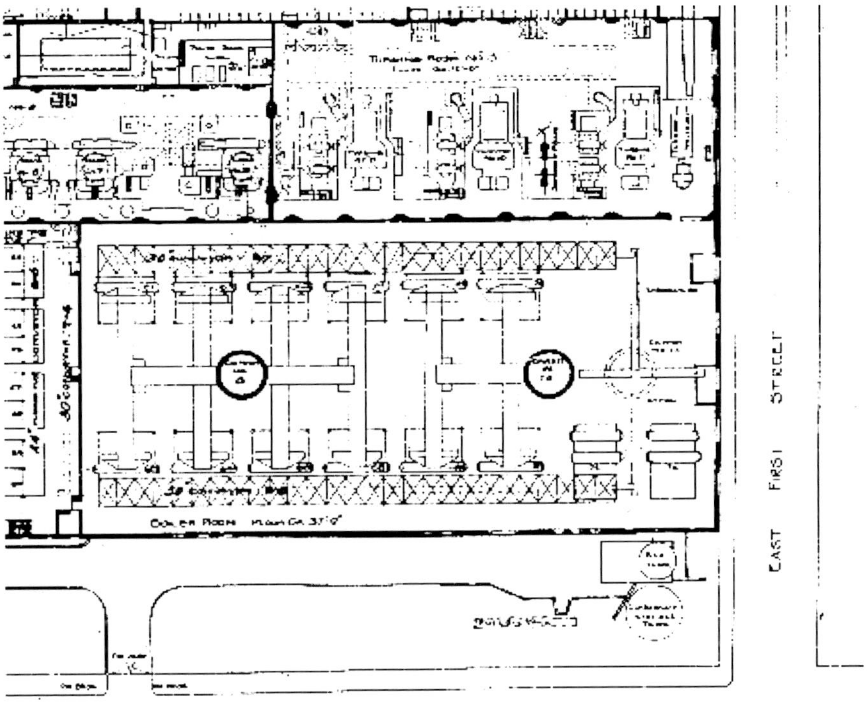

Figure 3-2 – Plan of Boiler Room 3 and Turbine Room 3 with turbines, boilers and chimneys built for the 300 psi steam expansion. Space has been set aside for future high pressure boilers #74 and #76.

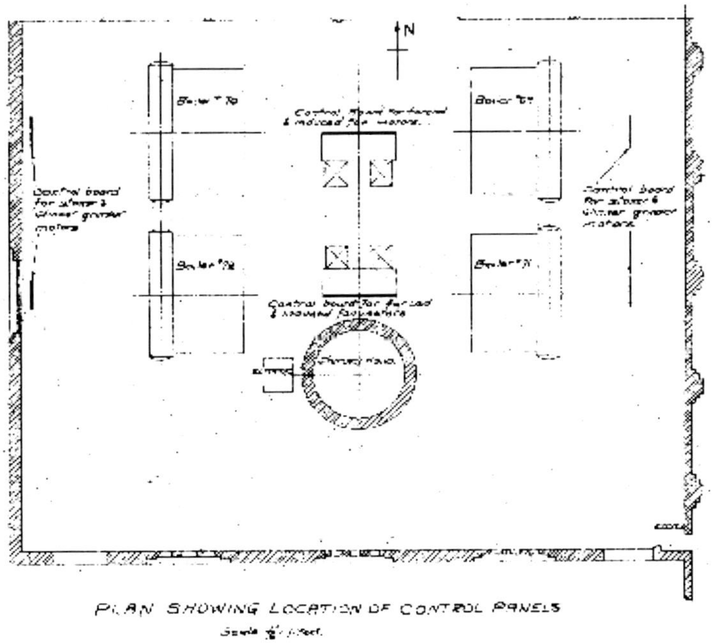

Figure 3-3 – Partial plan showing chimney C10 and local control boards for boilers #69, 70, 71 and #72. Boilers had motorized induced and forced draft fans, stokers and clinker grinders.

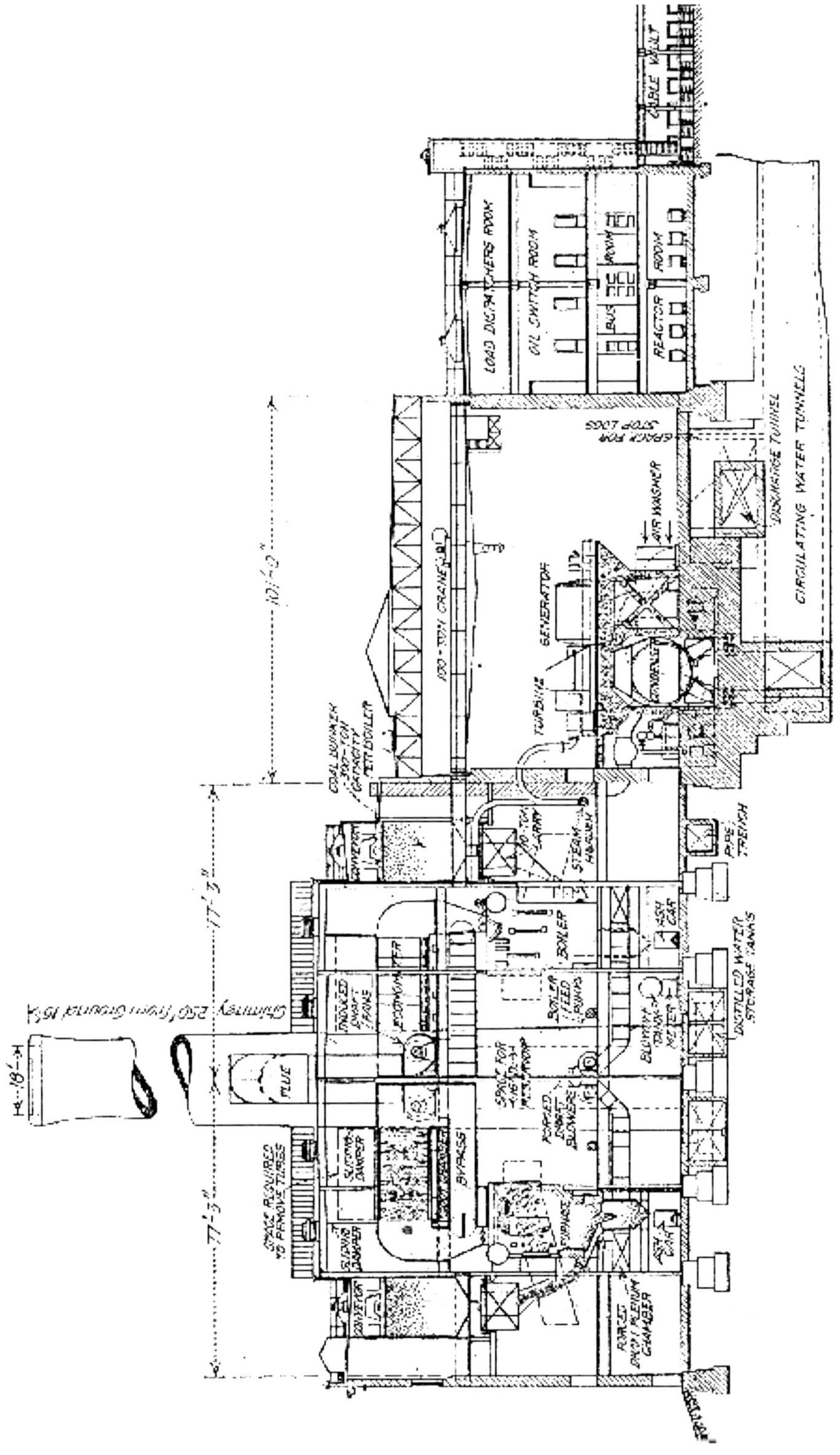

Figure 3-4 – Cross section of the 300 psi power station.

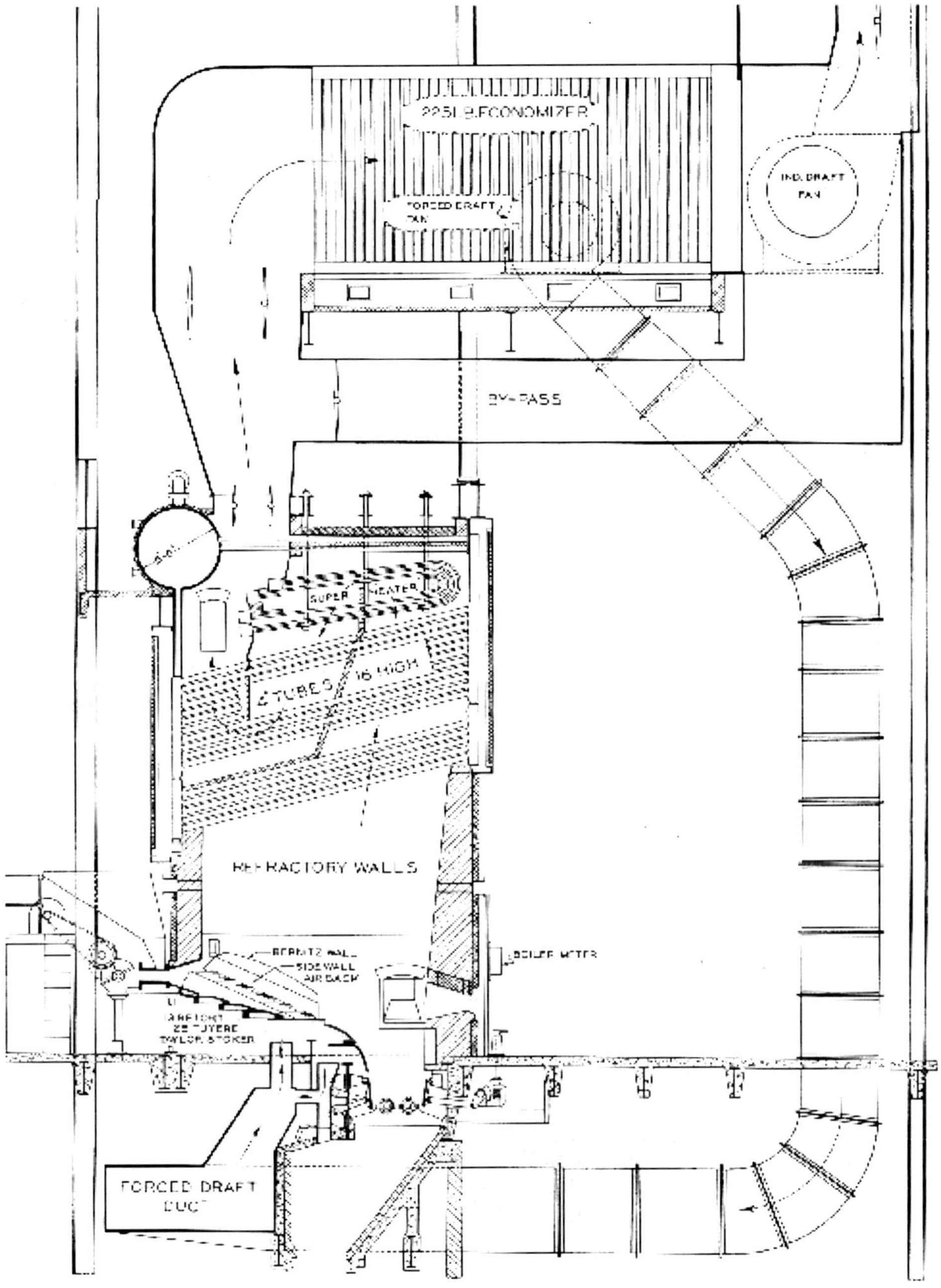

CROSS SECTION OF 14.579 SQ.FT. 300 LB. BOILER
L STREET STATION

Figure 3-5 – Cross section of the 300 psi B&W boiler.

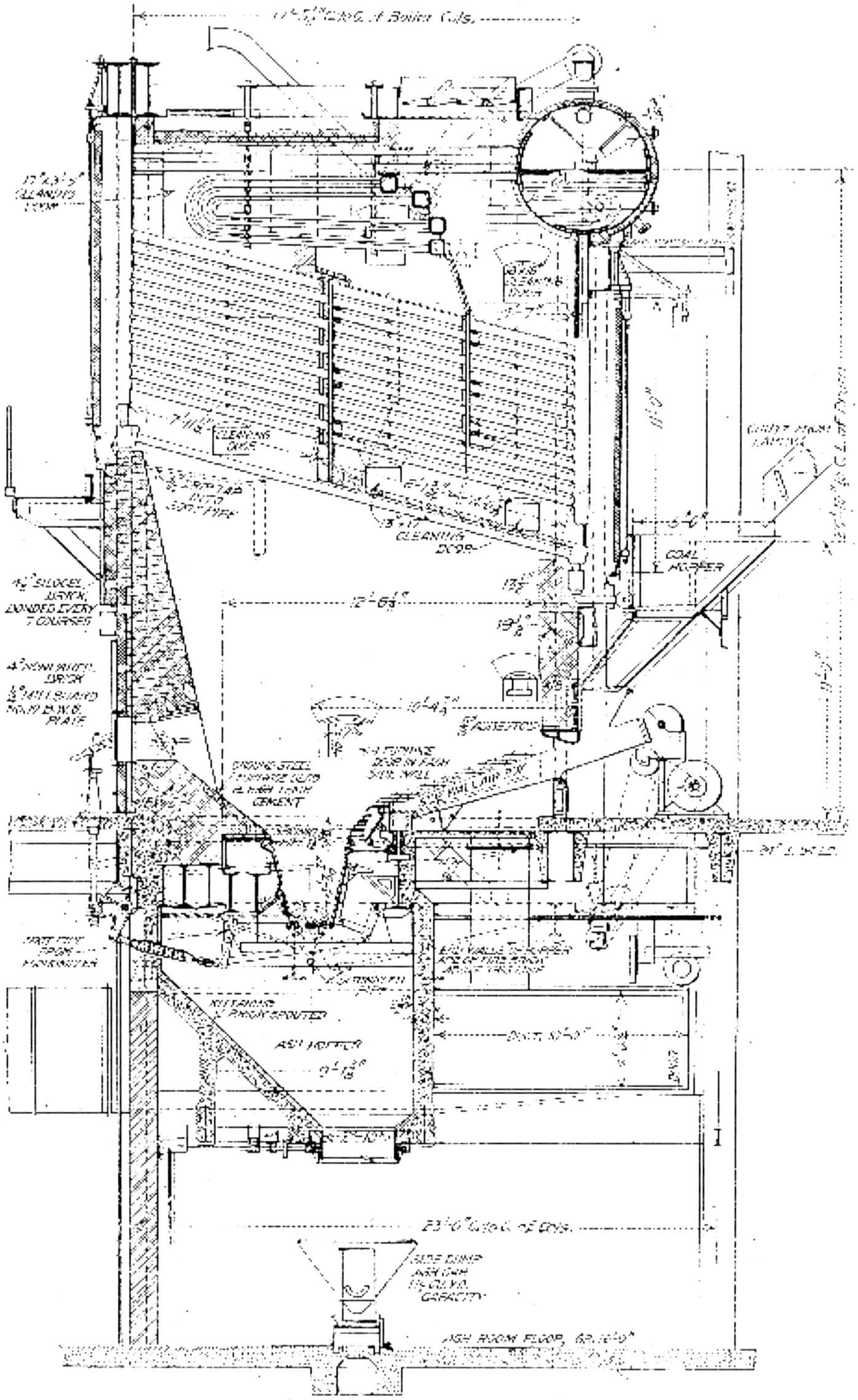

Figure 3-6 – A look inside one of the 300 psi cross drum boiler showing superheaters, refractory, stokers, clinker grinders and ash hopper.

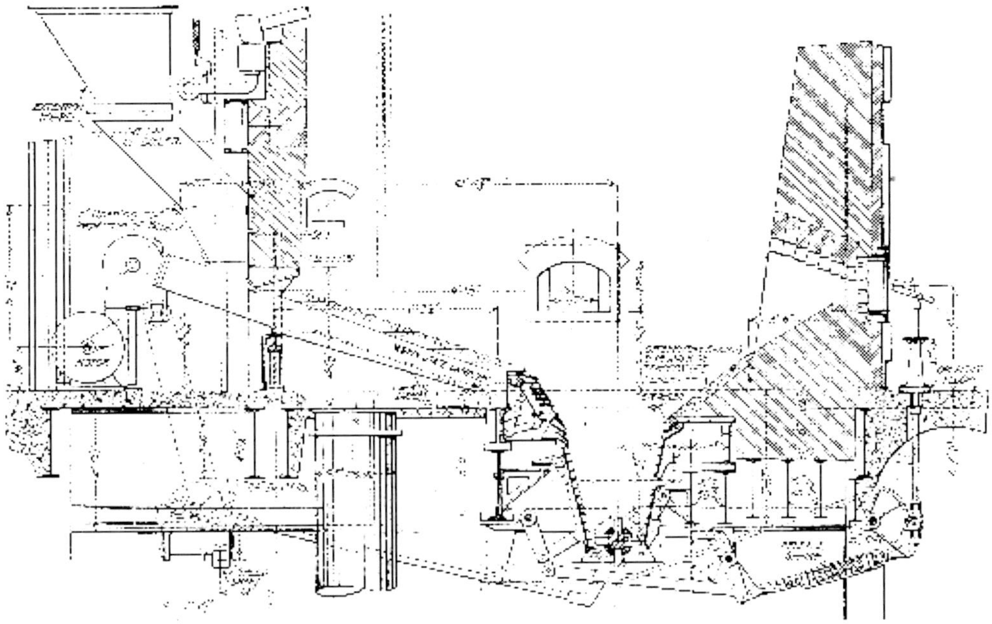

Figure 3-7 – The station was equipped with the latest clinker grinders, specifically made for large boilers. These 1919 automatic underfeed stokers and controls were manufactured by Westinghouse.

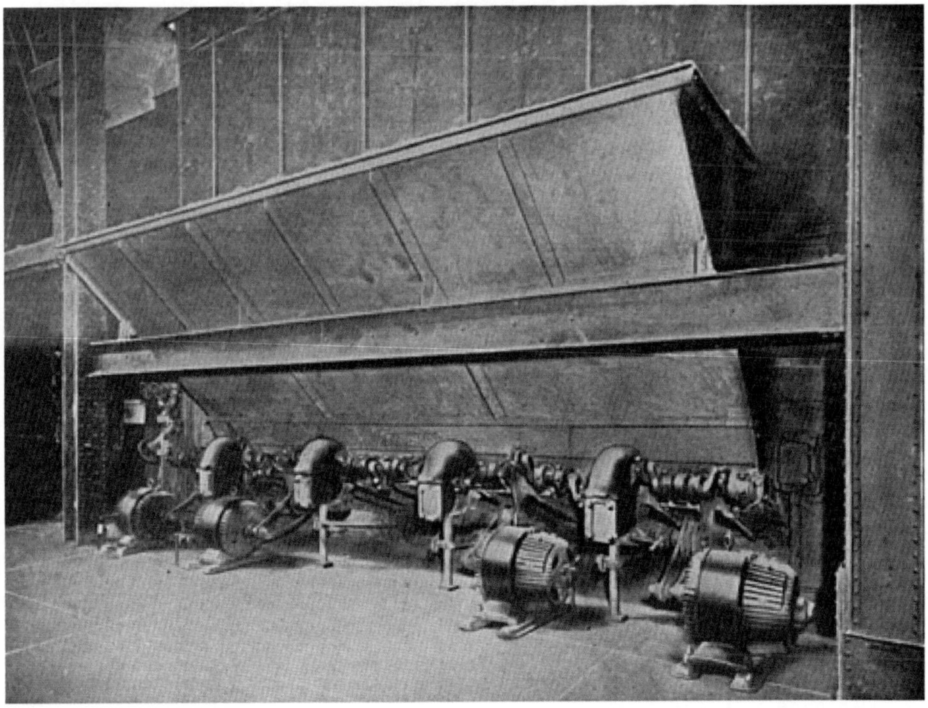

Figure 3-8 – The firing aisle where coal hoppers and stokers were located.

Figure 3-9 – This was the operating or control side of the boiler. Notice the task lights used by the night shift.

Figure 3-10 – As demands for steam increased, stokers became a critical component in the steam plant. Boston Edison experimented with all major stoker manufacturers, like the Riley self-dumping underfeed stoker shown here. Notice the complexity of moving parts pushing and advancing the burning coal to the center of the furnace.

Figure 3-11 – The inside of a cross drum boiler under construction in another state. L Street boilers were identical.

Figure 3-12 – All steam and exhaust piping were prefabricated and erected by Lumsden & Van Stone of Boston. Note the cluster of motor operated valves with their local control panels.

Figure 3-13 – Turbine #9 with operators. May 24, 1920

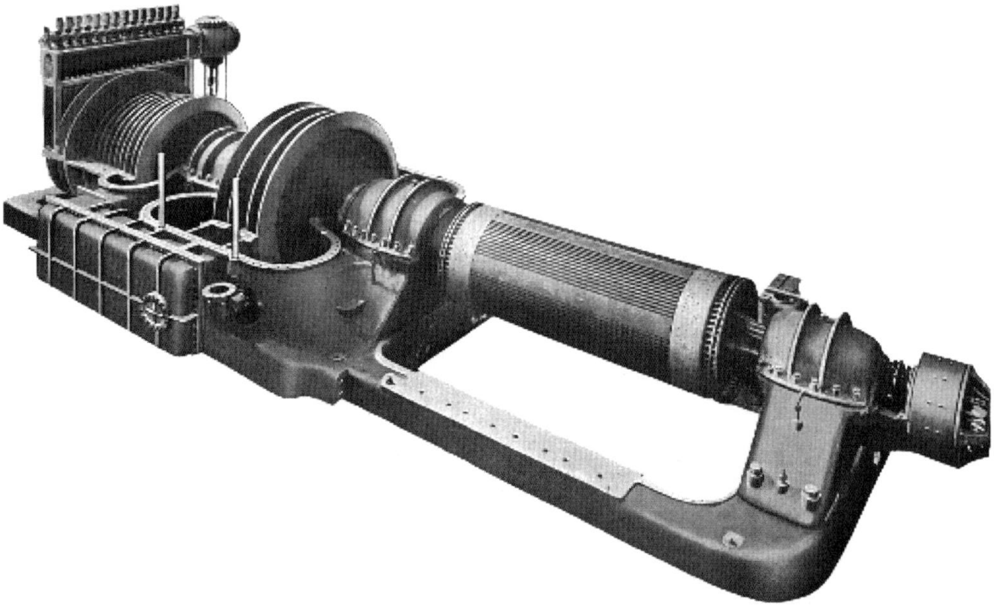

Figure 3-14 – This is what the inside of a 1916 GE 35,000 kw turbine generator looked like. The turbine rotor is shown assembled in the lower half of its casing. Steam admission valves are on the left, generator armature on the right.

Figure 3-15 – Turbine #10 was a 35,000 kw GE horizontal turbine generator, seen here on its pedestal. This unit was one of the first generators with a shaft-mounted exciter. March 21, 1921.

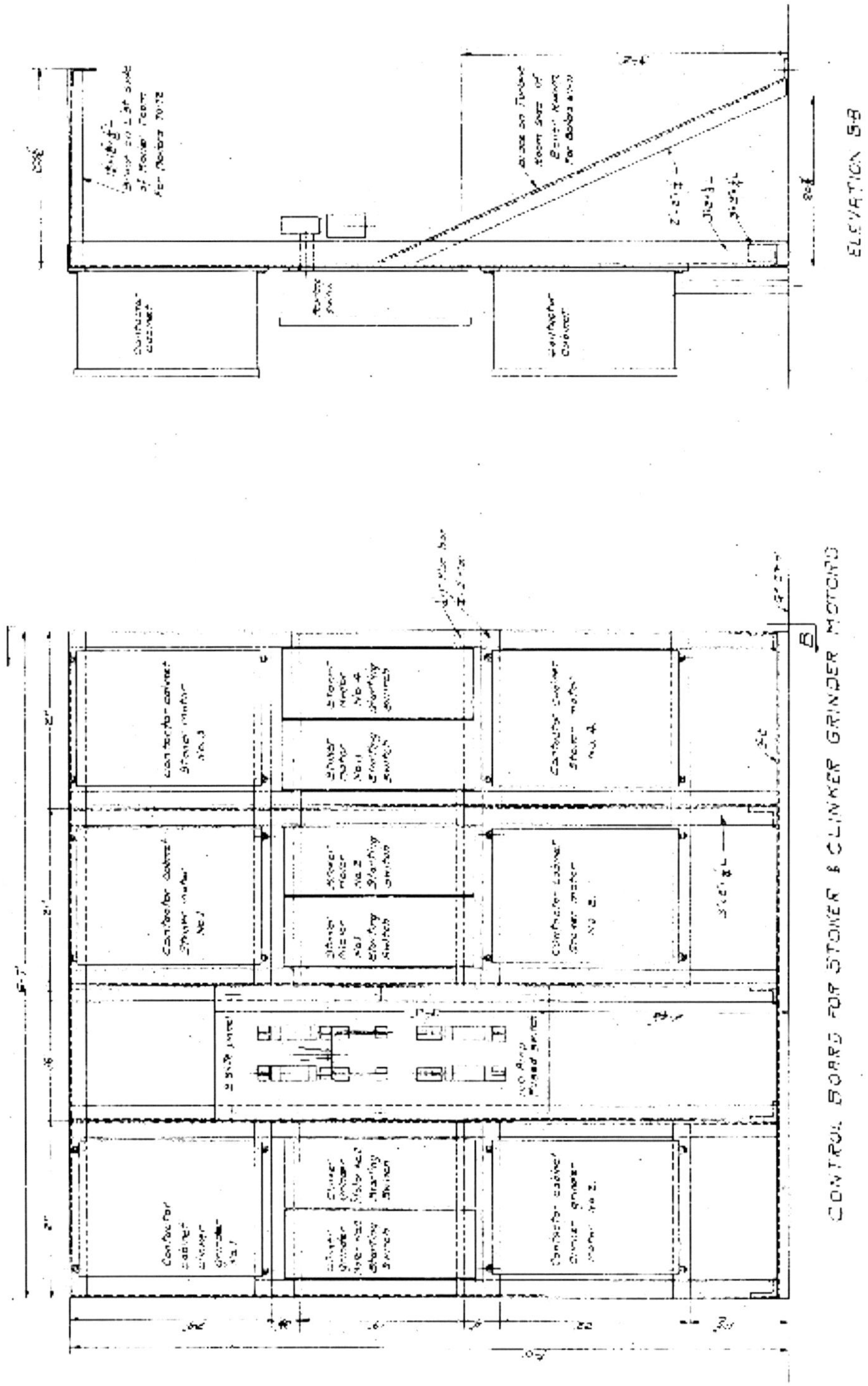

Figure 3-16 – A control board for stoker and clinker grinder motors. Metal enclosures for motor contactors, relays and control, were surface mounted to slate surfaces. Three-phase motors were 550 volt AC.

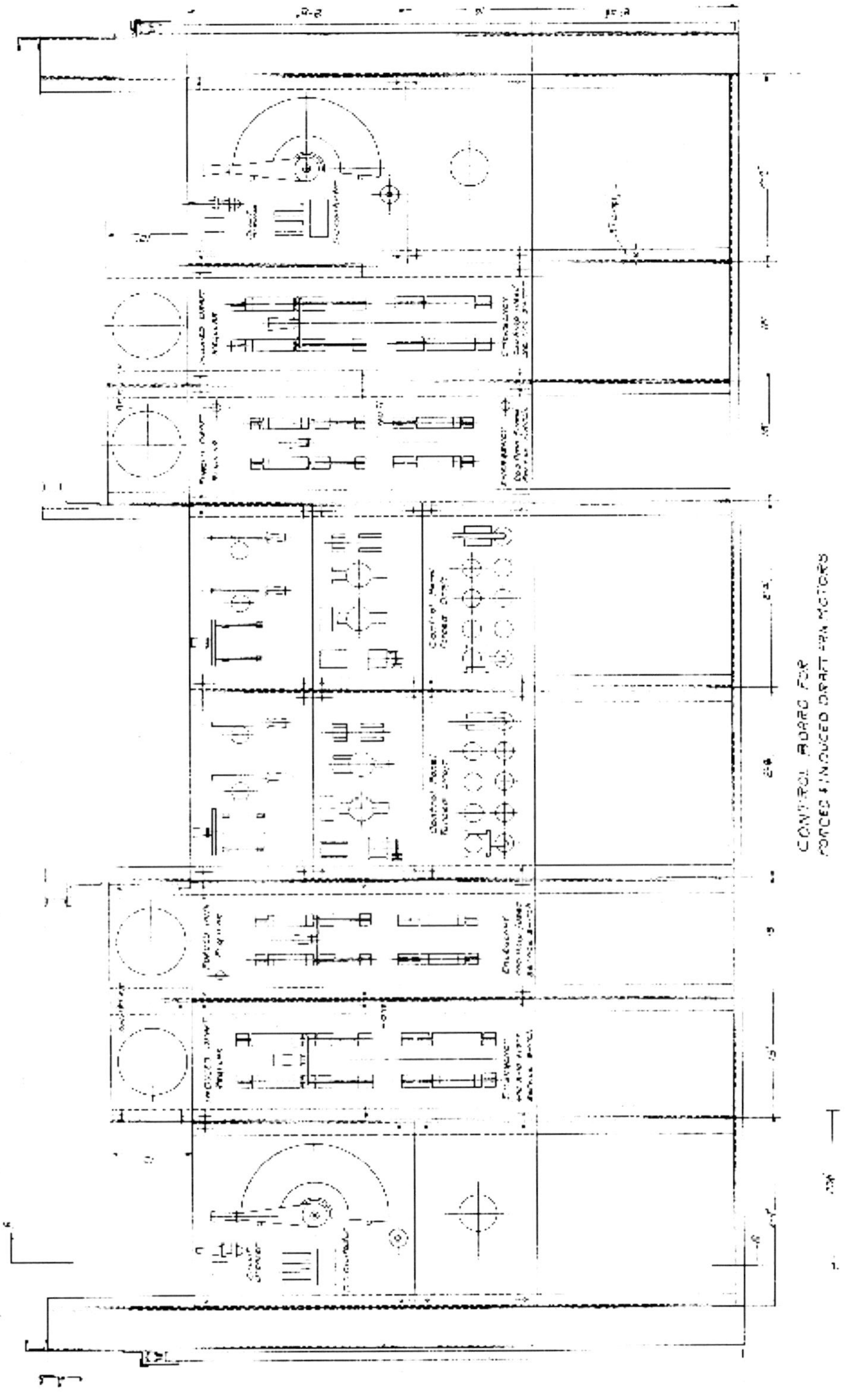

Figure 3-17 – Motor controls on slate board for force draft fans and induced draft fans.

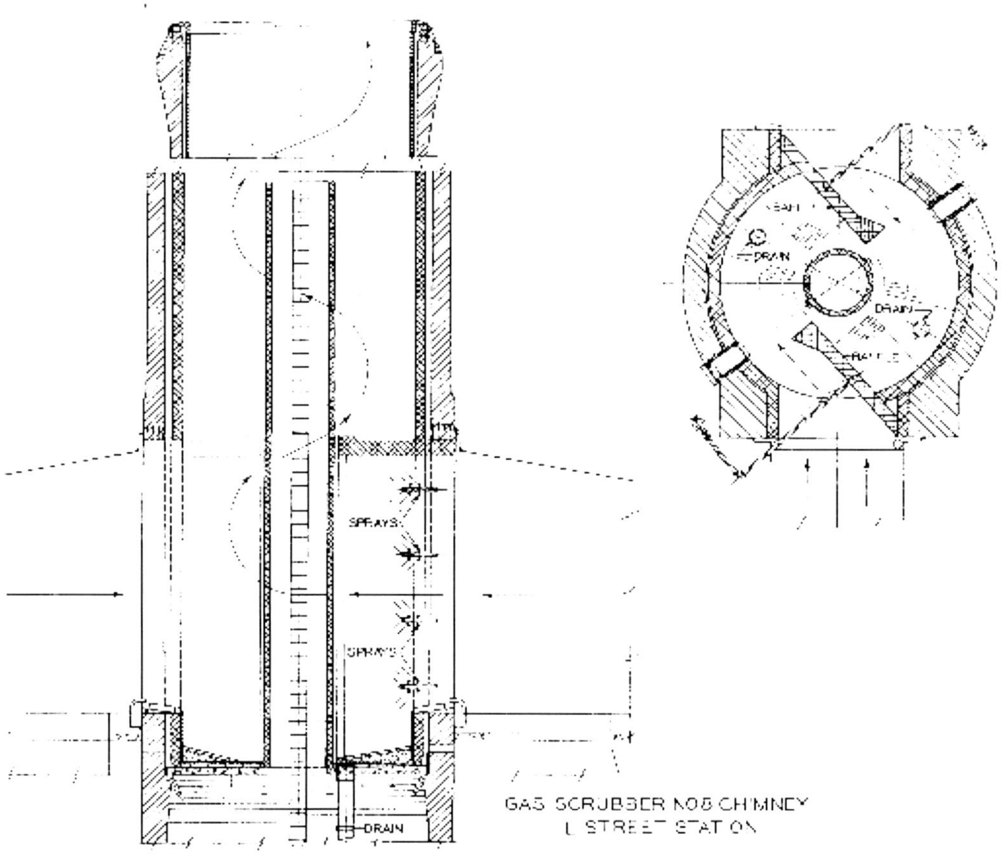

Figure 3-18 – In 1937, pioneering gas scrubbers were installed in chimneys C8 and C10.

Construction Photographs – Period 1917 to 1919

Exterior of the new Turbine Room 3. September 10, 1918.

Interior of the new Turbine Room 3. September 10, 1918.

Upper level of the boiler house with economizer in the background. Elevator is on the right, next to the chimney. April 4, 1918.

Extension of boiler house, side view, showing progress on boiler #61, boiler tubes and Westinghouse stoker grate. March 5, 1918.

Turbine parts hauled into Turbine Room 3. July 30, 1918.

General front view of progress on boiler #61 with Westinghouse stoker. March 5, 1919.

Front wall of turbine room. August 9, 1922.

200 lbs. to 300 lbs. reducing stations. August 25, 1918.

Work on boiler room extension, L Street wall. November 6, 1917.

Operating room panels for turbines #9 and #10, showing instruments, meters, control switches, and pilot lights. Note the selsyn rotary switches for sending orders to the turbine floor. March 3, 1921.

Work on boiler room extension with chimney C8 almost complete. July 3, 1917.

Part 4 | Historical development of the steam power generating system associated with chimney C12.

Historical Background

Edison Electric Illuminating Company of Boston was the first electric utility in this country to build a high pressure 1,200 psi steam generating plant. That took place at their Edgar Station in Weymouth, Mass. Later in 1926, the company installed two high pressure boilers and one turbine for operations at 1,400 psi pressure and temperature of 750 °F. This power installation was recognized as a landmark project by The American Society of Mechanical Engineers (ASME).

After several years of operation, EEI had clearly demonstrated the financial gain in going to such a pressure. In spite of higher first costs, improved fuel economy quickly offset the increased fixed charges. The experience gained at Weymouth gave the company impetus to apply high pressure steam in their next expansion program at L Street later in 1939.

In the twenties, there had been an increased tendency among power engineers to improve the steam cycle. This involved the development of satisfactory equipment. At the same time, considerable thought had been devoted to obtaining greater power capability from existing equipment without large financial investments. One promising field, explored by Boston Edison, was superimposing high pressure steam on top of their existing low pressure steam. This was referred to as "topping."

Topping was supplying high pressure steam to a turbine and exhausting directly into a lower steam header. 1,400 psi steam would be supplied to a turbine exhausting into 200 or 300 psi steam headers. Condensers were not needed.

This design was promising because of its large existing steam supply. Many of the original 200 psi boilers had exceeded their life expectancy. With the introduction of topping boilers and pressure reducing stations, obsolete boilers could be retired.

Plans for a high pressure steam expansion at L Street were initiated in December 1937, when the company purchased two 1,400 psi topping boilers and a high pressure turbine.

Equipment Description

The first phase of this project was completed in 1939 and consisted of the following equipment:

- Boiler #74 and #76: manufactured by B&W. Each boiler had a capacity of 406,000 lbs. steam at 1,400 psi, with a superheater. Each boiler was provided with pneumatic combustion controls furnished by Bailey Meter Company. This would be the first complete draft and combustion controls introduced at L Street Station. Pneumatic instruments and controls were housed in a modern control room located in Boiler Room 3.
- Because of the high firing rate required, new boilers used powdered coal or pulverized coal as fuel. The problem of excessive temperatures resulting from firing pulverized fuel, caused damage to bricks and solid refractory. This led boiler manufacturers to develop water-cooled furnace wall construction. Pulverized coal was introduced at L Street at that time. Oil burners and oil burning equipment were installed for use as an alternate fuel.
- The new high pressure boilers employed welded drums, soot blowers, air heaters, and furnace water walls. Automatic combustion controls became an important part of the boiler package because there were just too many variables to be manually managed.
- The turbine generator T #1201 was a high pressure GE unit, rated 25,000 kw.

The second phase of this expansion project was completed in 1942, consisting of the following:

- Boiler #75: Manufactured by Combustion Engineering Company, having a capacity of 542,000 lbs. steam at 1,400 psi and 910 °F, with superheat at 914 °F.
- The turbine generator T #1202 was a high pressure Westinghouse unit, rated 25,000 kw.

Steam System

The best way to understand L Street's steam distribution system is to refer to the Steam Flow Diagram available in one of the attached figures.

In general, steam flowed from the boiler drum, through the superheater, then to the high-pressure 1,400 psi piping. The 1,400 psi steam was normally admitted to one or both of the turbines, exhausting directly into a 300 psi steam header. This header was connected to the existing piping system. This steam then became available for the 300 psi horizontal turbines. Exhaust steam also supplemented the 200 psi steam by going through pressure reducing valves. Pressure reducing valves, check valves and isolation valves could be used to by-pass the turbines, directing steam flow to low pressure steam headers.

Frequency Converter

A significant system development took place in 1950 with the addition of a 50,000 kw frequency changer or frequency converter. This machine made possible the interchange of power between Boston Edison and the Metropolitan Transit Authority, predecessor of the Massachusetts Bay Transit Authority (MBTA).

Frequency converters were motor-generator sets often used to connect two power systems together with different frequencies, to exchange energy. Unlike ordinary motor-generators, this type of service required that the frequency converter be reversible, that is, either machine becoming the driving motor. Both machines had to be synchronous machines.

At the time, the MTA had a large 25 cycle power plant located next door to L Street, generating electricity for traction power. The frequency converter was installed in the old engine generator room and went in service in 1950.

1952 Inventory of Station Equipment and Overall Performance

Nameplate capacity ____________ 249,750 kw

Gross capacity ________________ 251,000 kw

Boilers:

- Twenty seven 200 psi, 537 °F. Installed 1904 – 1911
- Twelve 300 psi, 622 °F. Installed 1919 – 1924
- Three 1,400 psi, 910 °F. Installed 1939 – 1942

Turbines nameplate capacity:

- Eight verticals, 200 psi. Total 96,000 kw. Installed 1904-1914
- Three horizontals, 300 psi. Total 105,000 kw. Installed 1919-22
- Two topping 1,200 psi. Total 48,750 kw. Installed 1939-1942

Frequency changer: 50,000 kw

- Net output ______________ 857,861,678 kwhr
- Net economy ____________ 15,766 BTU per kwhr
- Coal field capacity ________ 56,000 short tons
- Oil storage ______________ 161,716 barrels
- Coal used ______________ 292,517 short tons
- Oil used ________________ 865,356 barrels
- Precipitators ____________ 300 and 1,400 psi boilers

L Street's System Contribution

L Street's contribution to Boston Edison's overall power system will be illustrated at the end of this section. The first chart presents a plot of fuel economy data taken from 1926 to 1951. It shows that the average quantity of fuel required to generate a kilowatt-hour of electricity gradually decreased during that time period. The second chart will show the contribution to the company's overall kilowatt generating capabilities during 1932 to 1956.

Postscript

The 300 psi boilers were shut down around 1973.

Turbines T #9, #10 & #11 remained in reserve until 1974.

The 1,400 psi steam system remained in service until approximately 1978, supplying steam to Boston customers.

Turbine #1201 stayed in service exhausting steam into the 300 psi until 1978.

Turbine #1202 ran till approximately 1974 then was dismantled and removed.

Photographs and Illustrations

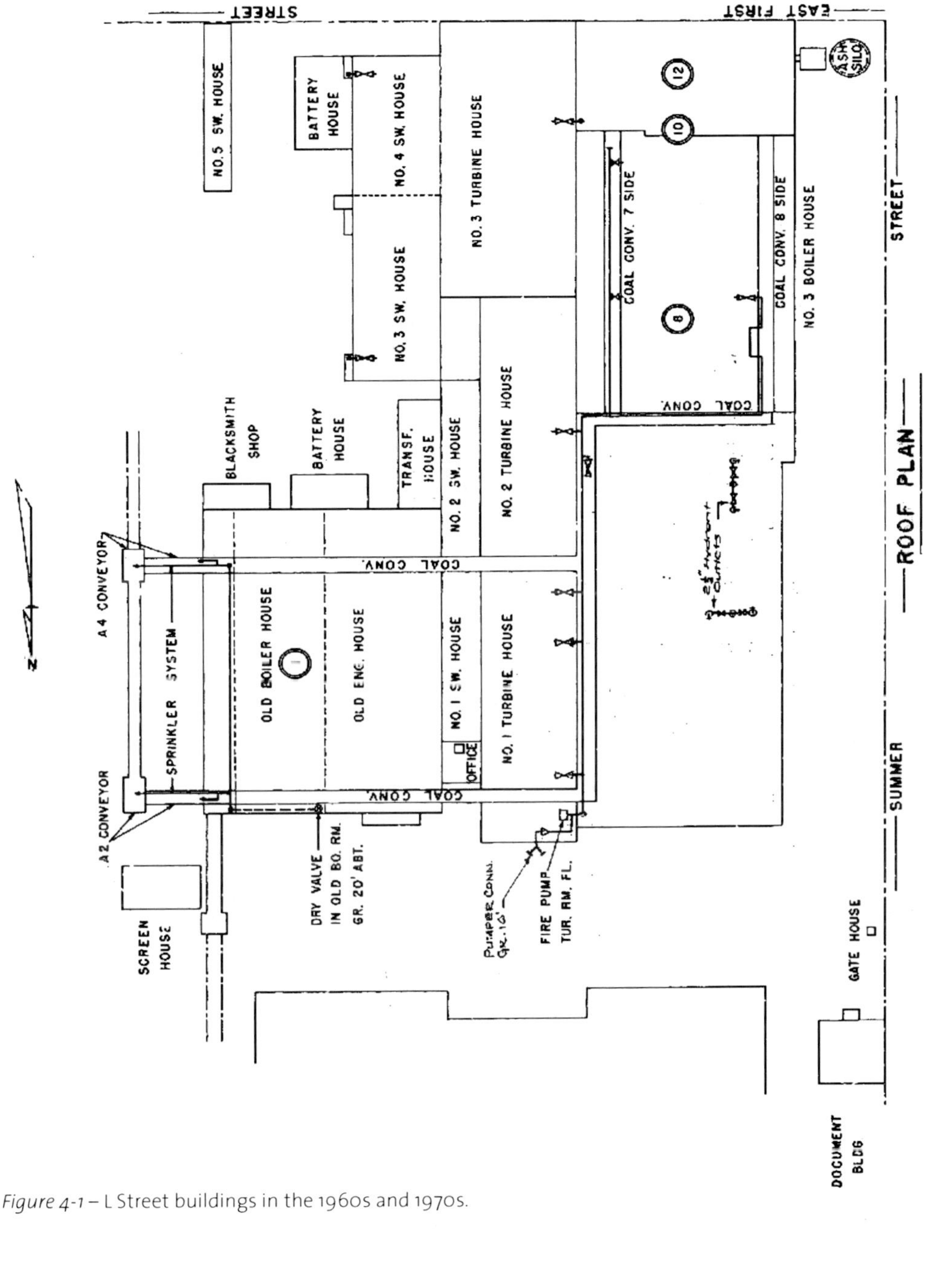

Figure 4-1 – L Street buildings in the 1960s and 1970s.

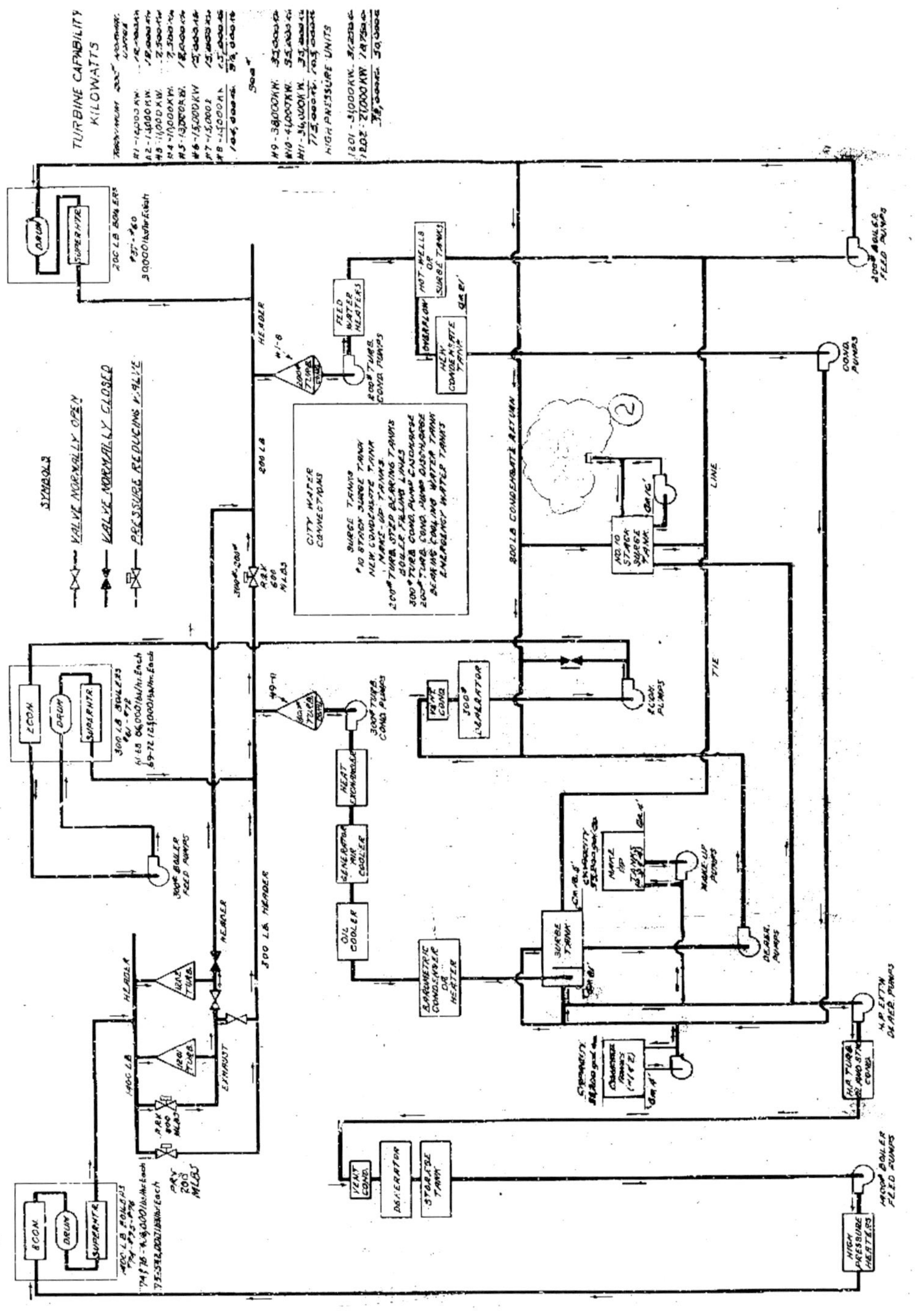

Figure 4-2 – Steam Flow Diagram issued by EEI in 1938.

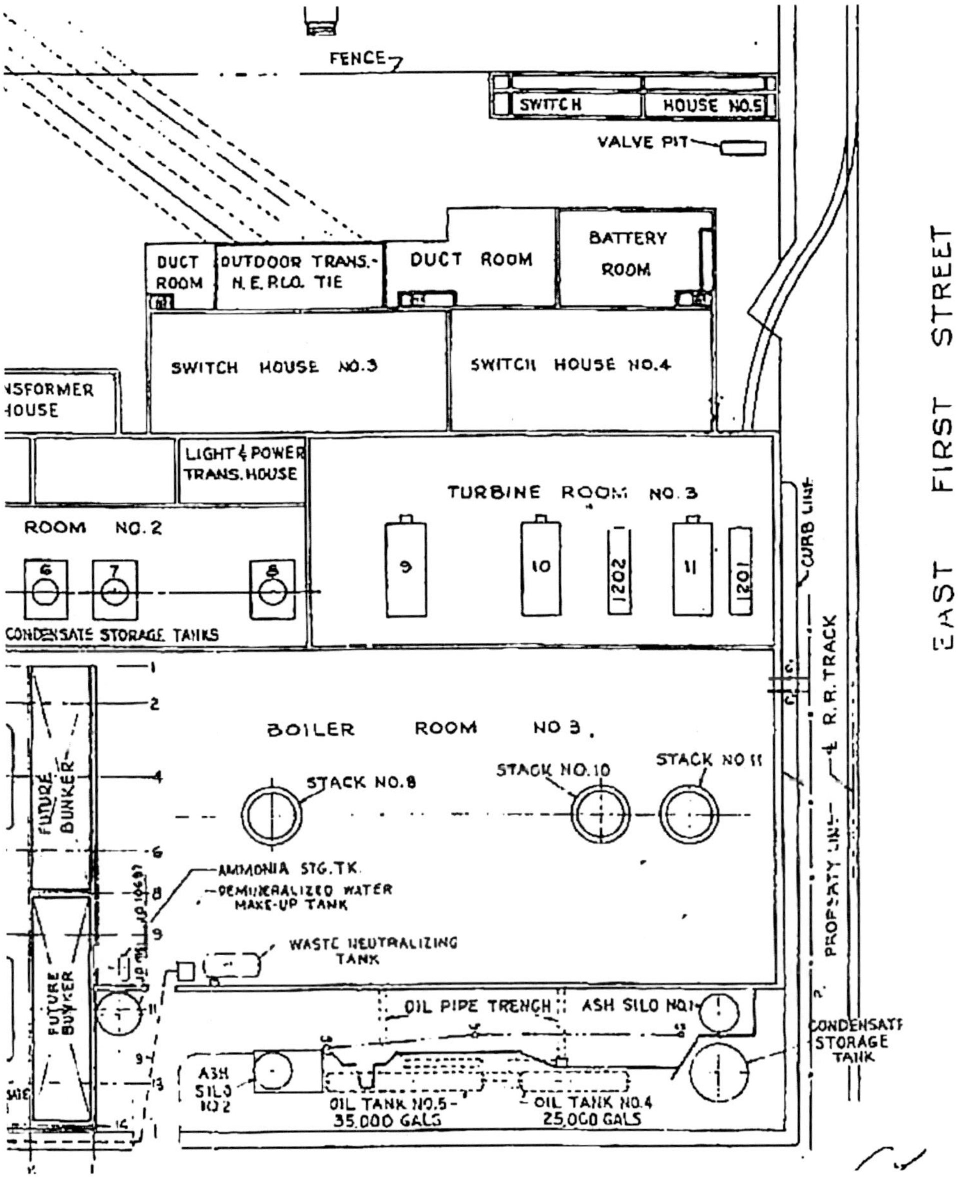

Figure 4-3 – Part of Stone & Webster Engineering Company's 1963 Plot Plan showing the arrangement of equipment in Boiler Room 3 and Turbine Room 3. Stack number 11 is labeled incorrectly—it should read number 12.

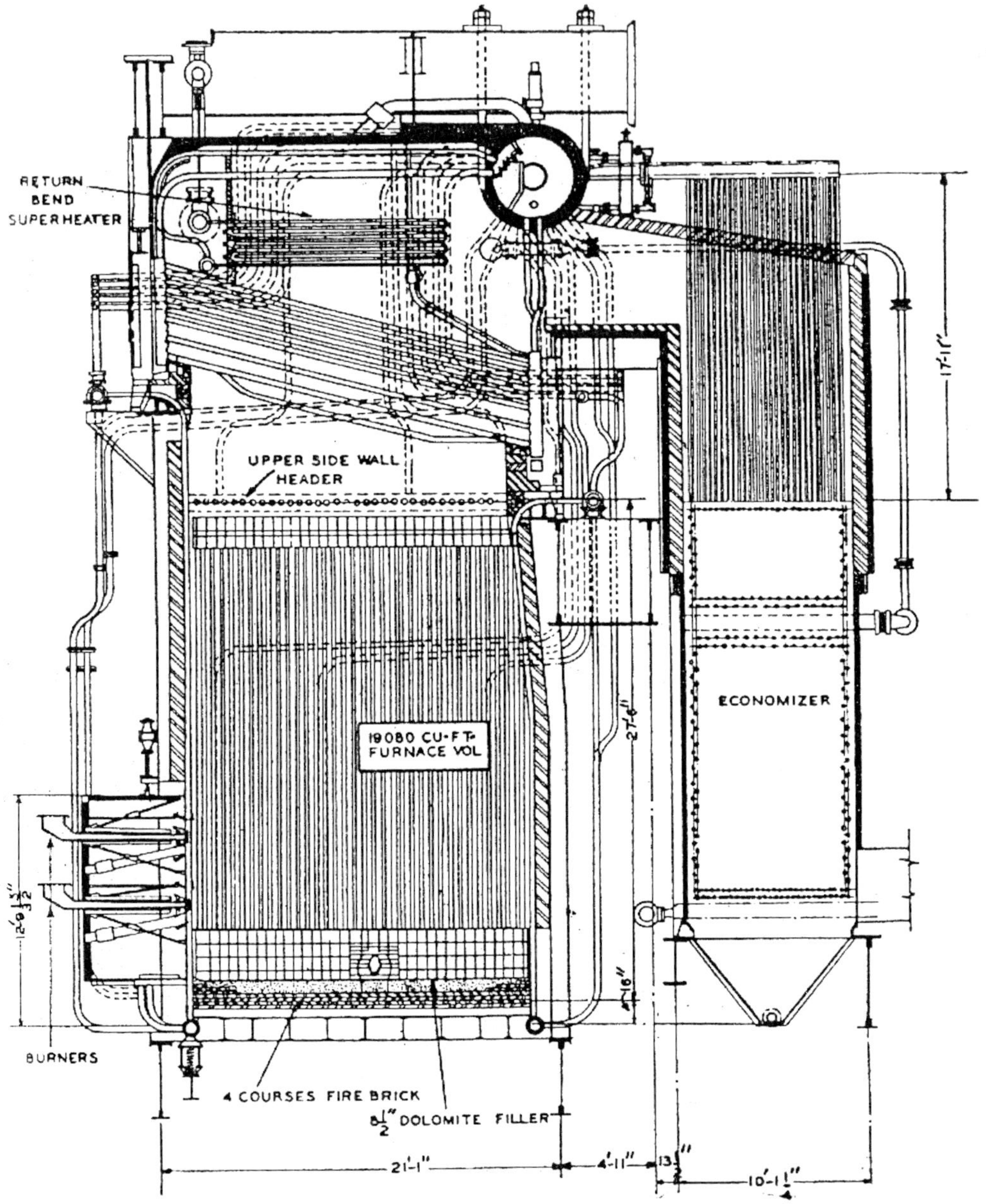

Figure 4-4 – Cross section of the 1,400 psi boiler manufactured by B&W. The design was highly compact, featuring a welded steel drum, economizer, and water-cooled furnace. The primary fuel was pulverized coal with provisions for diesel fuel oil as alternate.

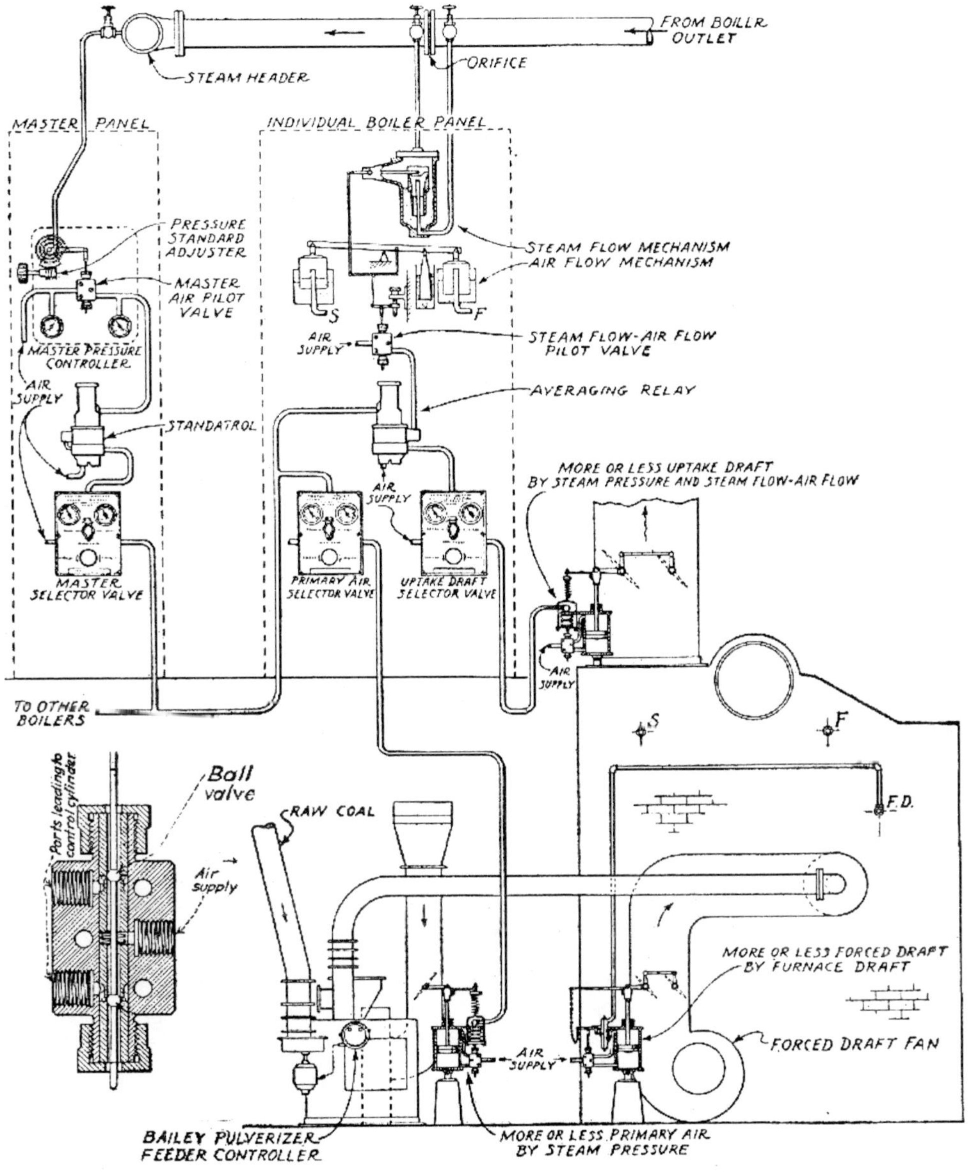

Figure 4-5 – This is a simplified Bailey Meter Company combustion control system for a pulverized coal fired boiler. Steam pressure is used to initiate the master control signal, which is transmitted through to other components to adjust or set fuel, air and other variables.

Figure 4-6 – A heavy piece of the frequency converter being hauled to the site through the rear entrance on East First Street. The building on the left is switchhouse No. 5. The battery house is on the right.

Figure 4-7 – The frequency converter was assembled in the old engine house of L Street Station.

Figure 4-8 – One of the two rotors being assembled.

Figure 4-9 – The frequency converter went in service in 1950. (Savaria 2006).

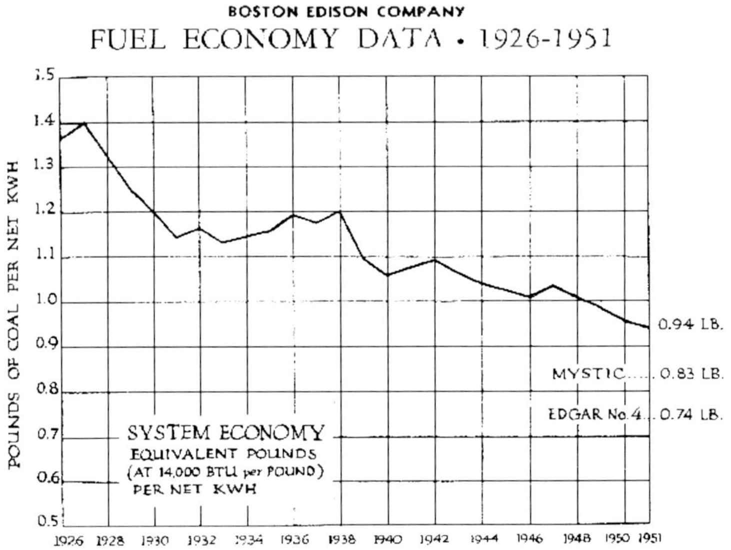

Figure 4-10 – Boston Edison System Fuel Economy Data: 1926 to 1951

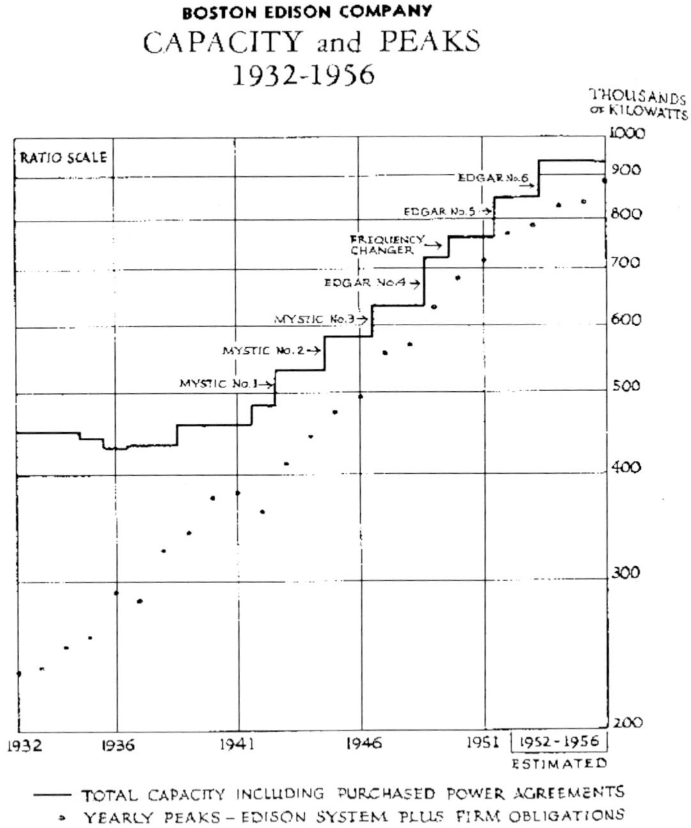

Figure 4-11 – Shows the affect of adding 25,000 kw (T 1201) in 1939; 24,000 kw (T 1202) in 1942; and 50,000 kw (frequency converter) in 1950.

Station Property Photographs – 1953

L Street Station at South Boston—capacity 249,750 kilowatts.

Section of turbine room at L Street Station showing three 35,000 kilowatt and one 18,750 kilowatt turbo-generating units.

Part 5 | A brief history of New Boston Generating Station.

Figure 5-0 – New Boston Generating Station in 2006. Sithe Energies Corporation acquired the facility in May 1998. It was subsequently sold to Exelon Power Corporation in 2002. Its candy-pink exterior finish is an acquired taste: it will surely grow on you if you look at it more than once. Part of the 1903 Edison Electric Illuminating Company brick building, with its white arched facade, is visible to the left of the new structure. (Savaria 2006).

A New Beginning

New Boston was a quantum leap in power engineering for Boston and New England. The word "new" in naming the station was a way to distinguish it from the older L Street power systems, many of which were a century old. L Street Station had been built using dozens of small boilers, and of small prime-movers with matching generators. Boston Edison decided to adopt the modern concept of one large boiler supplying one large turbine-generator, a marked departure from earlier days. Another important design feature was the use of once-through subcritical steam boilers, rather than conventional steam drums.

Engineering

Early in the 1960s, Boston Edison contracted with Stone and Webster Engineering Company to investigate the feasibility of constructing a new power plant on the existing site along Summer Street. Project engineering was performed on a fixed-price basis instead of their standard reimbursable financing. The project was launched and construction was completed very quickly.

Construction was especially difficult due to existing operating conditions and cramped quarters, which were close to a busy street. It was necessary to dismantle and remove a great deal of old equipment without interfering with the operation of the rest of the station. Boilers, coal bunkers, duct work, fans, stokers and stacks were taken away and then the roof was removed. Some of the existing brick walls were reused. During excavation for Unit 2, construction crews found human remains and a skull. A check of land records indicated that a cemetery existed at the site in late 1800. Boston Edison did the right thing by purchasing a cemetery plot to bury the remains. It's unclear whether the workforce was given the day off to attend the funeral.

By 1965, the first New Boston unit in service was 50 percent larger than any other operating in New England. When the second unit was completed in 1967, New Boston became the largest generating power station operating in a six-state area.

Startup and Commercial Operation Dates

Unit 1: began commercial operation August 11, 1965.

Unit 2: began commercial operation July 17, 1967.

General Building Plan

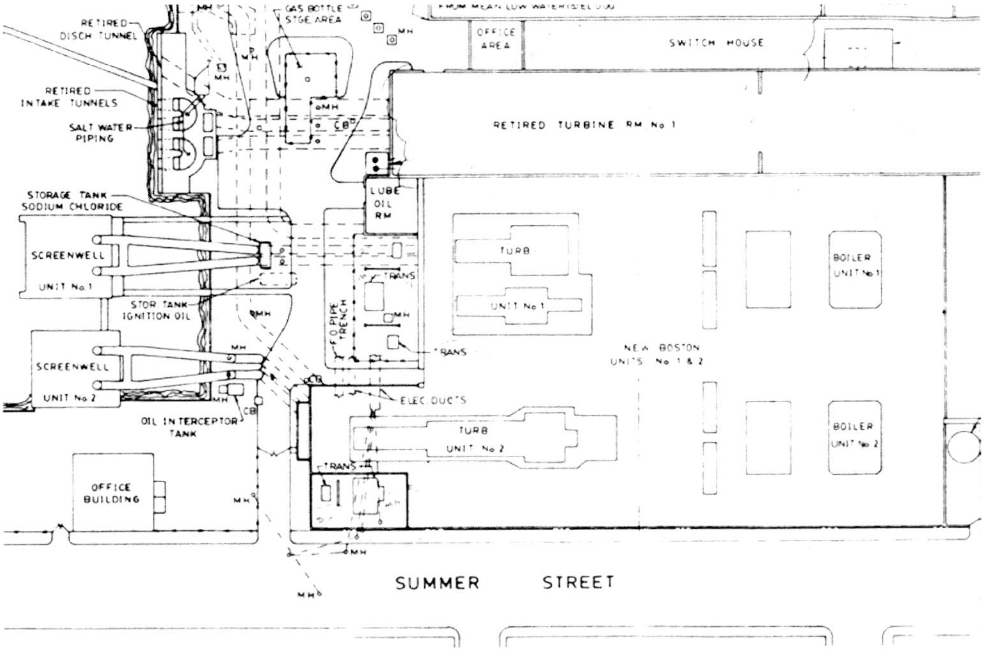

Figure 5-1 – The two New Boston units were built on space once occupied by boilers #13 through #60 and chimneys C2, C4, and C6. Units were arranged parallel to Summer Street with generators located at the north end of the building. Unit condensers receive cooling water from the Reserve Channel.

Plant Description

Boilers:

— Two Babcock & Wilcox Universal-Pressure boilers with a capacity of 2,640,000 lbs/hr of steam at 2400 psi at 1000 °F. Designated as UP-17 and UP-38 for Units 1 and 2 respectively.

Turbines:

— Unit #1 is a Westinghouse two cylinder cross-compound, double flow, condensing reheat turbine generator; the speed of the high pressure turbine is 3,600 rpm, steam inlet pressure 2,400 psi, temperature 1,000 °F. Speed of the low pressure turbine is 1,800 rpm. The generators are inner-cooled by hydrogen under pressure as high as 60 psi.

— Unit #2 is a General Electric tandem-compound turbine generator; operating at 3,600 rpm, high pressure turbine.

The boilers are Babcock & Wilcox Universal-Pressure (UP) type. Steam is produced at 2,400 psi and 1,000 °F and reheat steam is also the same temperature. The UP boiler operates on the concept of a long single tube into which water is pumped at one end and fire is applied in the center, which results in steam leaving the other end at the right pressure and temperature. The firing rate determines the quality and condition of the steam output.

Each unit has two force draft fans and two gas recirculating fans. The recirculating fans take their suction from the economizer hopper section and supply cool gas to the recirculating ducts or to the tempering ducts. Recirculating gas is admitted into the ash pit and flows up the furnace wall, forming a blanket of insulation, which retards the heat absorption in the walls. The tempering gas is admitted into ducts on each side of the boiler, which carries it up to the top of the furnace. It enters just below the secondary superheater section, forming a blanket of gas across the top of the furnace, causing more heat to be absorbed in the boiler walls. The dampers are controlled from a central control board. Each boiler has 2 sets of 12 guns. Each set of guns is arranged in three levels of four guns each. Originally, oil was supplied to the burners at a pressure range of 200 to 1,050 psi. Three fuel oil transfer pumps pumped oil from the oil tank farm.

Each unit has two boiler feed pumps. The auxiliary boiler feed pump is the startup pump, and is driven by a 4,500 hp motor. It will deliver up to 1 million pounds per hour. The main boiler feed pump is driven by the turbine shaft. The condensate is taken from the hotwell by two condensate pumps and pumped through a jet condenser, a gland seal condenser, to three condensate demineralizers. The boiler feed pump takes its suction from the deaerator and supplies feedwater through a series of high pressure heaters into the economizer. From the economizer, feedwater passes through the radiant sections of the boiler, then through various superheater sections. From the high pressure turbine rotor, steam flows back to the reheater. From there, the steam flows to the intermediate rotor, over to the low pressure turbine, returning to the condenser to begin the cycle all over again.

Extraction steam is taken off at seven points. Condensers are required to receive a large amount of condensate and steam during startup operations. Each unit has a dedicated closed cooling water system. Cooling water is pumped through two salt water heat exchangers, which are supplied by the circulating water intake lines and discharge into the outfall line into the channel.

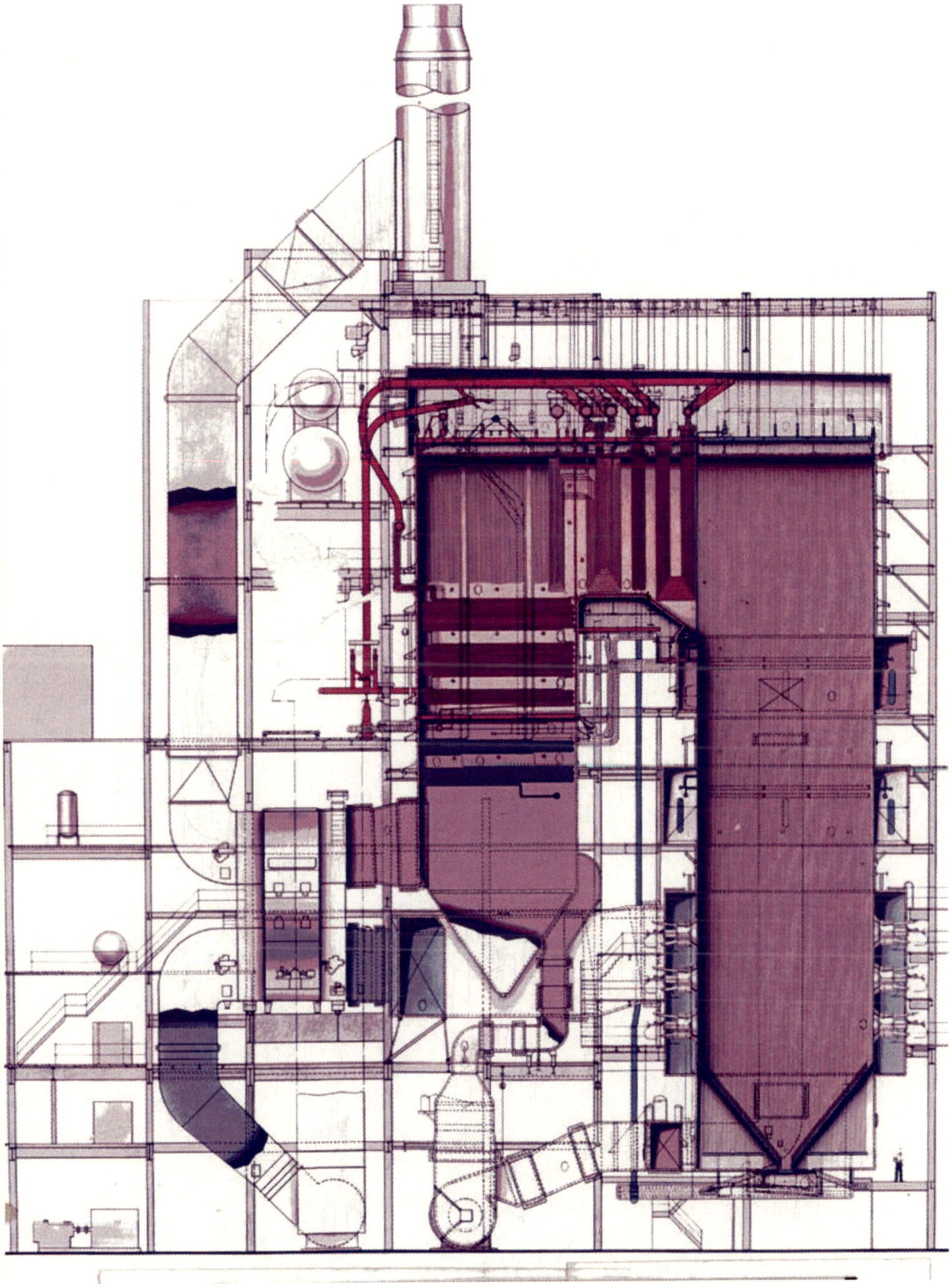

Figure 5-2 – Cross section of Babcock & Wilcox Universal-Pressure once-through subcritical boiler. Major design features include: water-wall cooling, absence of any steam drum, three tiers of oil/gas burners, two horizontal-flow air-heaters, economizer, primary and secondary superheaters, reheater unit, two gas recirculation fans and two force draft fans. Two induced draft fans were added in 1993.

Figure 5-3 – Unit 1 photographed in 1965 during startup tests just before being placed in service. (Courtesy of Boston Public Library, Prints Department).

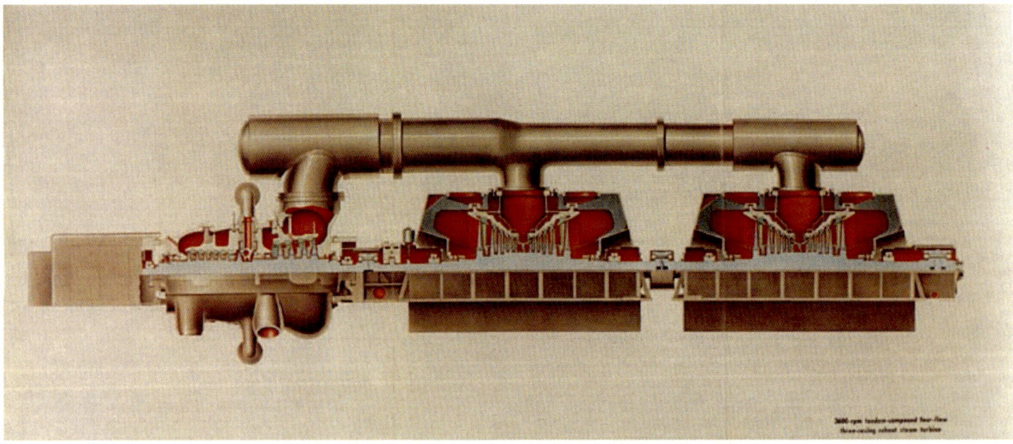

Figure 5-4 – Cross section of Unit 2 General Electric steam turbine.

Fuel

Number 6 diesel fuel oil was the primary fuel delivered at 800 psi to mechanical atomizers. Natural gas became the secondary fuel following the 1985 natural gas conversion project. In order to reduce sulfur and particulate emissions, it was adopted as primary fuel April 1, 1995. The fuel oil system was decommissioned in 2004. Fuel tanks were emptied. Obsolete oil pipes were evacuated and "pigged."

Tall Chimney and Balanced Draft Modifications

The units' boiler draft design was changed in 1993 when two induced draft fans were added to each boiler. The change reduced boiler furnace pressure from positive (15 inches water column) to negative (five inches water column). The net effect eliminated flue gas leakage from the furnace as the boiler now operated at a slight vacuum. The four exhaust stacks (two per unit) were replaced with two higher stacks (one per unit) as part of the flue gas design change. Existing burners were replaced with low NO_X burners.

Figure 5-5 – New Boston was originally designed with two flue stacks for each of the two boilers. The 1993 Tall Chimney and Balanced Draft Project completely changed all that. This photograph was taken July 1967. (Courtesy of Boston Public Library, Prints Department).

Figure 5-6 – As the generation of electricity became more complicated, requirements for more sophisticated instrumentation and integrated controls became more and more evident. The original Bailey Pneumatic 721 system was later upgraded to Bailey digital controls. Here is the original Information Control Center in 1965. (Courtesy of Boston Public Library, Prints Department).

Figure 5-7 – Maintenance crew removing the low pressure generator rotor for inspection and overhaul in 1995.

Power Plant Explosion

On October 2, 2002, newspaper headlines reported an explosion inside the plant caused by hydrogen-cooled generator Unit 2. The initiating cause was the unintentional closure of breakers with the unit unsynchronized to the grid, turning the generator into a motor, a condition known as "motorizing the generator." This caused the generator to accelerate uncontrollably from zero to 2,200 rpm, severely damaging the generator and causing the fire. Fire Commissioner Paul Christian said, "It's absolutely one of the worst firefights you can have. There has been a terrific amount of heat and a tremendous amount of smoke." Firefighters had to go up stairways in the windowless building in blinding smoke. Flames shot up 50 feet inside the turbine building. The fire department quickly upgraded their initial response to a 9-alarm blaze. Over 20 employees were evacuated from the plant, which suffered $10 million in damages. There were no fatalities.

Operating Statistics

	Service Hours New Boston	
Year	**Unit 1**	**Unit 2**
1965	2,930	
1966	7,008	
1967	6,888	3,057
1968	6,515	7,668
1969	6,989	7,590
1970	7,372	5,334
1971	7,421	7,453
1972	6,148	7,929
1973	6,614	7,182
1974	5,351	6,599
1975	5,398	6,724
1976	4,890	6,568
1977	6,270	5,900
1978	7,336	6,245
1979	6,461	5,718
1980	6,326	8,002
1981	4,675	7,297
1982	7,701	6,688
1983	6,811	7,324
1984	4,922	6,824
1985	6,901	4,578
1986	6,720	7,331
1987	5,813	7,642
1988	6,575	6,756
1989	7,307	8,108
1990	5,396	7,867
1991	7,709	4,914
1992	8,160	7,809
1993	5,660	7,046
1994	6,333	7,517
1995	4,969	8,222
1996	7,865	4,454
1997	5,704	8,537
1998	6,621	7,532
1999	3,855	3,201
2000	2,157	3,053
2001	1,990	4,303
2002	1,876	2,392
2003	1,173	
2004	992	
2005	1,454	
2006	1,338	
Total	**230,594**	**231,364**

	Annual Net Generation (mWh) New Boston	
Year	**Unit 1**	**Unit 2**
1965	803,936	
1966	2,487,066	
1967	2,149,949	1,220,000
1968	2,144,203	2,449,630
1969	2,230,757	2,374,067
1970	2,445,726	1,704,003
1971	2,124,690	2,429,360
1972	1,928,393	2,575,240
1973	2,172,099	2,244,840
1974	1,700,118	2,172,437
1975	1,746,790	2,162,894
1976	1,467,850	2,017,120
1977	1,938,668	1,813,540
1978	2,326,102	1,961,482
1979	1,994,522	1,722,480
1980	1,892,795	2,410,569
1981	1,313,435	2,170,440
1982	2,275,320	1,822,458
1983	2,082,350	2,135,950
1984	1,485,270	2,027,610
1985	2,150,330	1,380,990
1986	2,110,549	2,201,689
1987	1,823,060	2,336,831
1988	2,064,940	2,114,140
1989	2,330,802	2,600,380
1990	1,714,562	2,329,251
1991	2,274,068	1,196,574
1992	2,200,669	1,957,912
1993	1,429,417	1,687,226
1994	1,504,998	1,631,308
1995	1,287,535	1,979,550
1996	1,842,050	1,035,714
1997	1,439,886	1,929,285
1998	1,192,709	1,412,271
1999	757,998	573,238
2000	350,780	459,662
2001	336,455	796,913
2002	311,530	456,406
2003	199,135	
2004	160,563	
2005	246,860	
2006	220,851	
Total	**66,659,786**	**65,493,460**

Plant Performance Capabilities

As far as the New England Power Pool (NEPOOL) is concerned, Units 1 and 2 are essentially the same, except for differences in heat rates. The following NEPOOL data is abbreviated and based on firing natural gas.

Unit 1:

— Normal Claimed Capability:____________ 350 MW net

— Reserve Claimed Capability:____________ 365 MW net

— Maximum Claimed Capability: _________ 380 MW net

— Response Rate: ________________________ manual: 6.0 MW per minute

________________________ automatic: 4.0 MW per minute

— Start-up Time from Hot Conditions: ____ 24 hours

— Start-up Time from Cold Conditions: ___ 16 hours

— Average Heat Rate at Winter Normal Net Capability:

________________________ 9,737 Btu/Net kwh at 350 MW.

Unit 2

— Normal Claimed Capability:____________ 350 MW net

— Reserve Claimed Capability:____________ 365 MW net

— Maximum Claimed Capability: _________ 380 MW net

— Response Rate: ________________________ manual: 6.0 MW per minute

________________________ automatic: 4.0 MW per minute

— Start-up Time from Hot Conditions: ____ 24 hours

— Start-up Time from Cold Conditions: ___ 16 hours

— Average Heat Rate at Winter Normal Net Capability:

________________________ 9,854 Btu/Net kwh at 350 MW.

Highest Annual Output

The highest annual output occurred in 1998, when 4.9 million megawatt hours net was reported, equating to a 73.6 percent capacity factor.

New Boston Station Employees—November 2006

General Manager

John Castro

Operations Manager

David O'Brien

Watch Engineers

Christopher Capen
William Dougherty
Francis Ferguson
James O'Hare
James Pugliese
William Wilkins

Central Control Operators

Louis Clark
Patrick Downey
Robert McLeod
Paul O'Brien
Paul Poirier
Thomas Roach
Robert Stickney
Stephen Swirka
Stephen Clark
1 Vacant

Equipment Operators

Robert Crawford
Anthony DeCouto
Kenneth Lawlor
Joseph Pugliese
Carol Sampson
1 Vacant

Chemical Control Operators

Mike Boguszewski
Bernard Moran
1 Vacant

Laborer

Cynthia Johnson

Maintenance Manager

John Ayvazian

Maintenance Supervisors

Dave Belanger
Donald Vaughan

Fossil Station Mechanics

James Cleaves
Robert Colton
Frank Declements
John Hart
Mike McSweeney
Scott Rusconi
Wayne Swanstrom
Keith Ware

Watch Electrician

Tom Laranjo

Instrument & Controls

Wilson Changeau
Dan O'Leary
Tom Webster
1 Vacant

Stockkeeper

Mary Lisa Freeman

Station Clerk

Lisa Jerome

Technical Manager

John Hughes

Environmental Engineer

MaryEllen Doherty

Financial Analyst

Melissa Naegelin

Purchasing Agent

George Triantafilopoulos

Recent Photographs by Savaria 2006

Unit 2.

Unit 1.

Operators inspecting Unit 1 fluid drive/main boiler feed pump coupling. Chris Capen, Watch Engineer (standing with flashlight); Ken Lawlor, Equipment Operator; Joe Pugliese, Equipment Operator (kneeling).

Operator completing surveillance task on digital control system display. Bob McLeod, Central Control Operator.

Unit 1 exciter end.

Control Room Operators preparing Unit 1 for start-up. *Foreground:* Lou Clark, Central Control Operator. *Back Row:* Bob McLeod, Central Control Operator; Ken Lawlor, Equipment Operator; Dave O'Brien, Operations Manager.

Jim O'Hare, Watch Engineer

Carol Sampson, Equipment Operator

Waste Water Treatment Training Session.

Back Row: Jim Pugliese, Watch Engineer; Maryellen Doherty, Environmental Coordinator.

Middle row: Tom Laranjo, Watch Electrician; Chris Capen, Watch Engineer;

Bob Stickney, Central Control Operator.

Front Row: Joe Pugliese, Equipment Operator; Dave O'Brien, Operations Manager;

Suzanne Savaria; Don Vaughan, Maintenance Supervisor.

[1] Tom Webster, I&C Technician; [2] Bernie Moran, Chemical Control Operator; [3] Jim Pugliese, Watch Engineer; [4] Jim O'Hare, Watch Engineer; [5] Lisa Freeman, Stock Keeper; [6] Joe Pugliese, Equipment Operator; [7] George Triantafilopoulos, Purchasing Agent; [8] Carol Sampson, Equipment Operator; [9] Frank DeClements, Fossil Station Mechanic; [10] Maryellen Doherty, Environmental Coordinator; [11] Mike McSweeney, Fossil Station Mechanic; [12] Ken Lawlor, Equipment Operator; [13] Chris Capen, Watch Engineer; [14] Lisa Jerome, Clerk; [15] Bob Stickney, Central Control Operator; [16] Jim Cleaves, Fossil Station Mechanic; [17] Dave O'Brien, Operations Manager; [18] John Ayvazian, Maintenance Manager; [19] Tom Laranjo, Watch Electrician; [20] Don Vaughan, Maintenance Supervisor; [21] Wayne Swanstrom, Fossil Station Mechanic; [22] Dan O'Leary, I&C Engineer.

Maryellen Doherty, Environmental Coordinator.

Lisa Jerome, Station Clerk.

Dave O'Brien, Operations Manager.

Steve Swirka, Central Control Operator.

George Triantafilopoulos, Purchasing Agent

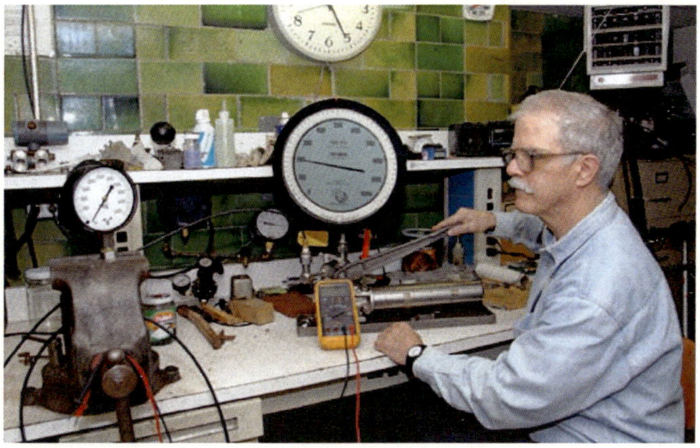

Dan O'Leary, I&C Engineer

Dave O'Brien, Operations Manager

Dan O'Leary, I&C Engineer; Ken Lawlor, Equipment Operator

Wayne Swanstrom, Fossil Station Mechanic; John Ayvazian, Maintenance Manager; Frank DeClements, Fossil Station Mechanic.

References

Books

Frank Koester, *Steam-Electric Power Plants*, 2nd edition. New York: D. Van Nostrand Co., 1910, pp. 31, 75, 400-405.

Joseph G. Workers and Thomas A. Peebles, *Mechanical Stokers*. New York: McGraw-Hill 1922, pp. 187-190.

T. Stevens and HM. Hobart, *Steam Turbine Engineering*. New York: The MacMillan Co., 1906, "Examples of Steam Turbine Plants", pp. 454-629.

J. Weingreen, *Electric Power Plant Engineering*. New York: McGraw-Hill, 1910, pp. 281, 283-285, 302.

Boston Electric Handbook, American Institute of Electrical Engineers. Boston, 1904, pp. 89-104.

David Neal Keller, *Stone & Webster 1889-1989, A Century of Integrity and Service*. Stone and Webster Incorporated, 1989, pp. 245-246.

Gustaf A Gaffert, *Steam Power Stations*. New York: McGraw Hill, 1937, pp. 429-431.

Electrical World 75th Anniversary Issue, *The Electrical Power Industry Past, Present and Future*. McGraw-Hill, May 21, 1949, pp. 61-134.

Published Articles

"The New Station of the Boston Electric Light Company", *The Electric Engineer*. November 24, 1898, pp. 505-508.

"Central Station of the Boston Electric Light Company", *Electrical World and Engineer*. May 27, 1899, pp. 693-698.

"Expansion of the Boston Edison System", *Electrical World and Engineer*. May 21, 1904, pp. 941-951.

I.E. Moultrop, "Modern Central Station Design as Exemplified by the New Turbo-Generator Station of the Edison Electric Illuminating Company of Boston". Proceedings of the AIEE Meeting, New York. January 27, 1905, pp. 19-33.

"The New Boston Edison Turbo-Generator Station", *Electric World and Engineer*. February 4, 1905, pp. 251-252.

"Boston Edison Equipment in a Nutshell", *Electrical World*. October 31, 1908, p. 937.

"Operation of the Boston Edison Company's L Street Station", *Electrical World*. May 27, 1909, pp. 1,283-1,289.

"The New Turbine Station of the Edison Electric Illuminating Company of Boston", *Power*. July 1905, pp. 389-400.

"Extension to L Street Station, Boston", *Power*. January 27, 1920, pp. 124-129.

"The Clinker Grinder in Modern Boiler Practice", *Power*. April 22, 1919, pp. 592-598.

"Flue Gases Scrubbed Out in Boston Edison Plant", *Electrical World*. August 28, 1937, pp. 37, 91.

EW Norris, "Operations of 1200-lb Pressure Generating Unit at the Weymouth Power Station, Edison Electric Illuminating Co., Boston". *General Electric Review*. November 1924, pp. 714-718.

References

Published Articles (cont.)

"Better Frequency Regulation". *General Electric Review.* Vol 21, 1918, p. 726.

Electrical World News Issue, "New England Utilities Order Two Generators", October 12, 1940.

AD Blake and PW Swain, "Fifty Years of Power", *Mechanical Engineering, ASME,* Vol 52. 1932, pp. 319-320.

Hosea Webster, "Steam Generators", *Mechanical Engineering, ASME,* Vol 52. 1932, pp. 359-365.

George A Orrok, "Central Stations", *Mechanical Engineering, ASME,* Vol 52. April 1930, pp. 324-334.

LL Elden, "The Boston Edison System in 1907". 23rd Annual Meeting of the Association of Edison Illuminating Companies, September 10-12, 1907.

"Presentation of the Accomplishments of the Edison Electric Illuminating Co. of Boston in Competition for the Charles A. Coffin Medal for 1926". (Booklet dated 1926 in Boston Edison Collection).

David B. Sicilia, "Boston Edison Centennial: History of the Boston Edison Company 1886-1986". (Manuscript in Boston Edison Collection).

Harvard University Baker Library, Boston Edison Collection

Station Repairs Diaries, 1904 to 1935.

EA series various journals.

DA series of statistics for Station 4-2.

DP series various journals.

Station Engineering Department Appropriations, 1897-1940.

Generating Department Fiscal Year Reports for 1907 and 1921.

Boston Edison Company Annual Reports.

Made in United States
North Haven, CT
30 August 2026

17689324R00064